Smaller Orders of Insects of the Galápagos Islands, Ecuador: Evolution, Ecology, and Diversity

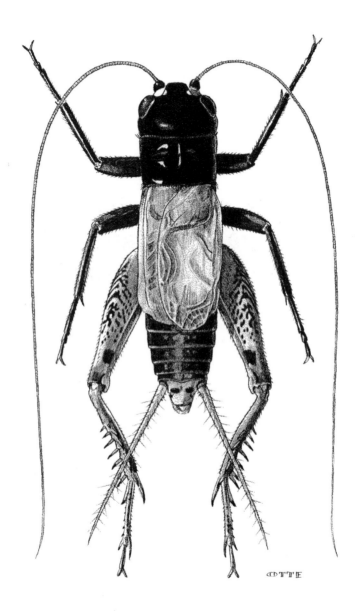

Gryllus darwini Otte and Peck from Darwin Island. One of a complex of six species of field crickets in the Galápagos Islands that have evolutionarily lost their second pair of wings and the ability to fly. Illustration by Dan Otte.

A PUBLICATION OF THE
NATIONAL RESEARCH COUNCIL OF CANADA
MONOGRAPH PUBLISHING PROGRAM

Smaller Orders of Insects of the Galápagos Islands, Ecuador: Evolution, Ecology, and Diversity

Stewart B. Peck

Department of Biology
Carleton University
Ottawa, Ontario
Canada K1S 5B6

NRC Research Press
Ottawa 2001

NRC Monograph Publishing Program

ISBN 0-660-18284-X
NRC No. 43262

Canadian Cataloguing in Publication Data

Peck, Stewart B.

Smaller orders of insects of the Galápagos Islands,
Ecuador: evolution, ecology, and diversity

Includes an abstract in French.
Includes bibliographical references.
Issued by the National Research Council of Canada.
ISBN 0-660-18284-X

1. Insects — Galapagos Islands 2. Insects — Evolution — Galapagos Islands.
I. National Research Council Canada. II. Title.

QL522.3E2P42 2000   595.7'09866'5   C00-980387-4

Correct citation for this publication:
   Peck, S.B. 2001. Smaller orders of insects of the Galápagos Islands, Ecuador: Evolution, Ecology, and Diversity. NRC Research Press, Ottawa, Ontario, Canada. 278 pp.

Contents

Prologue

"The natural history of these [Galápagos] islands is eminently curious, and well deserves attention. Most of the organic productions are aboriginal creations, found nowhere else; there is even a difference between the inhabitants of the different islands; yet all show a marked relationship with those of America, though separated from the continent by an open space of ocean, between 500 and 600 miles in width. The archipelago is a little world within itself, or rather a satellite attached to America, whence it has derived a few stray colonists, and has received the general character of its indigenous productions. Considering the small size of these islands, we feel the more astonished at the number of their aboriginal beings, and at their confined range. Seeing every height crowned with its crater, and the boundaries of most of the lava-streams still distinct, we are led to believe that within a period, geologically recent, the unbroken ocean was here spread out. Hence, both in space and time, we seem to be brought somewhere near to that great fact - that mystery of mysteries - the first appearance of new beings on this earth."

"I took great pains in collecting the insects [of the Galálapagos], but, excepting Tierra del Fuego, I never saw in this respect so poor a country. Even in the upper and damp region I produced very few, excepting some minute Diptera and Hymenoptera, mostly of common mundane forms. As before remarked, the insects, for a tropical region, are of very small size and dull colours. Of beetles I collected twenty-five species (excluding a *Dermestes* and *Corynetes* [Necrobia], imported wherever a ship touches); of these, two belong to the Harpalidae, two to the Hydrophilidae, nine to three families of Heteromera, and the remaining twelve to as many different families. This circumstance of insects (and I may add plants), where few in number, belonging to many different families is, I believe, very general. Mr. Waterhouse, who has published an account of the insects of this archipelago, and to whom I am indebted for the above details, informs me that there are several new genera; and that of the genera not new, one or two are American, and the rest of mundane distribution. With the exception of a wood-feeding *Apate* [Bostrichidae] and of one or probably two water-beetles from the American continent, all the species appear to be new."

C. Darwin, 1845, *Voyage of the Beagle*

Abstract/Résumé

The Galápagos Islands of Ecuador are world famous for their unique plants and animals, and the hints these gave to Charles Darwin in forming his ideas on evolution. They are the world's least altered set of tropical oceanic islands. The oldest have been available for colonization by land plants and animals for about 3–4 million years. At least 1850 species of insects and 350 species of other terrestrial arthropods are now known to occur on the islands.

This book is a synthesis of both previously published information and abundant new data derived from extensive recent field studies on Galápagos insects. The dynamics and patterns of the evolution, ecology, and distribution of the entire insect fauna are presented in general. The core of the book is an account of the 495 species of insects in the smaller orders (excluding Phthiraptera, Coleoptera, Diptera, Lepidoptera, and Hymenoptera), with detailed information on their distribution and bionomics This work will serve entomologists, island ecologists, evolutionary biologists, and conservation scientists as a framework to further the study and protection of the insects, the fauna and flora in general, and the islands themselves. It is a contribution to a comparative global understanding of the origin, evolution, and ecological structuring of oceanic island insect faunas.

Les Îles Galápagos de l'Équateur sont réputées dans le monde entier pour leur faune et leur flore qui ont inspiré la théorie de l'évolution à Charles Darwin. Il s'agit des îles d'océans tropicaux les plus intactes du monde. La plus ancienne île a commencé à accueillir végétaux et animaux terrestres il y a 3 ou 4 millions d'années. On recense sur les îles au moins 1850 espèces d'insectes et 350 espèces d'arthropodes terrestres.

Cet ouvrage constitue une synthèse de renseignements publiés antérieurement ainsi que de nombreuses nouvelles données résultant de récentes études de grande envergure sur les insectes des Îles Galapagos. On y présente une vue d'ensemble de la dynamique et des tendances de l'évolution, de l'écologie et de la répartition de toute la faune entomologique. Le corps de l'ouvrage est un relevé des 495 espèces d'insectes d'ordres mineurs (à l'exception des Phthiraptères, Coléoptères, Diptères, Lépidoptères et les Hyménoptères), offrant de l'information détaillée sur la répartition et la bionomie. Pour les entomologistes, les écologistes spécialisés en écologie insulaire, les biologistes évolutionnistes et les scientifiques de conservation, l'ouvrage s'avérera un outil d'approfondissement des connaissances, de protection des insectes, de la faune et de la flore en général de même que de préservation des Îles. Il contribuera à une compréhension comparative de l'origine, de l'évolution et de la structuration écologique des faunes d'insectes d'îles océaniques.

Summary

The Galápagos archipelago is composed of 19 islands larger than 1 km^2, with a total area of 7882 km^2. They are the world's only remaining tropical oceanic archipelago which is little altered by humans. The oldest islands have been available for terrestrial colonization for about 3-4 myr. At least 1850 species of insects and 350 species of other terrestrial arthropods are now known to occur on the islands. The total insect fauna is presently known to be composed of 292+ adventive species, 823+ indigenous species, and 735+ endemic species. The native species (indigenous and endemics combined) represent a rate of species accumulation of about one per 2250 years (through successful colonization plus speciation through about 3.5 million years). The main limitations to the establishment of new colonists seem to be the semi-arid climatic conditions of the large, lowland areas of the islands and the prevalence of young, unweathered lava substrates. The isolation of the islands by a distance of about 1000 km from the coast of South America is of lesser importance in the accumulation of the fauna, because other islands of similar size and isolation elsewhere in the world, but which are more humid, have more insect species. Aerial and sea-surface transport were the major modes by which insect colonists reached the islands. Sea-surface colonists included some short-winged or flightless species, and even some eyeless soil-inhabiting species. The principal source-area of the colonists ranged from Pacific coastal Mexico and Central America to northern South America. There are relatively few (24) endemic insect genera, and less than half the native species are endemic. Generic endemism and species multiplication is only partly correlated to secondary loss of flight wings (and of flight ability) after colonization. There are some monophyletic species swarms in flightless groups, but most colonization has not been followed by much species multiplication; the mean for the native insect fauna is about 1.13 species per colonizing ancestor. In the smaller orders, most species are generalist feeders on plant fluids, or are scavenging generalists, and none seem to have co-evolved with the endemic vegetation. There seems not to be an insect assemblage specialized for life on the young lava fields. The most dramatic evolutionary changes are in the many species that have become eyeless after occupying deep soil, basalt rock cracks, and cave habitats. There is a significant positive relationship of insect diversity with island area, elevation, and ecological complexity, but not island age. The complete taxonomic composition of the total insect fauna is probably now 90-95% sampled and sorted to species, but additional taxonomic work is needed to describe or to name the unidentified species. Few detailed field studies have been conducted on the ecological or evolutionary biology of the insects and there is much opportunity for additional study on the islands and for comparison with continental insect faunas.

This book is a synthesis of all previously published information on Galápagos insects and the new data derived from the field studies of my research teams. From these I have summarized the dynamics and patterns of the evolution and distribution of the entire insect fauna in general and the detailed ecological patterns of the smaller orders in particular. The body of this book is an annotated list of the 495 species of the smaller orders of insects now known to occur in the Galápagos Archipelago, with detailed information on their distribution and bionomics. Undescribed species are indicated when appropriate, but no new taxa are named here. Species level accounts form chapters for each of the apterygote, exopterygote (except Phthiraptera) and smaller endopterygote orders of insects. Later works will present similar summaries for the remaining orders: the ectoparasitic Phthiraptera lice and the four large endopterygote orders of insects (Coleoptera, Diptera, Lepidoptera, and Hymenoptera). I hope this work will help serve future entomologists, island ecologists, evolutionary biologists, and Galápagos National Park scientists and officials as a data base and as a framework to further the study and protection of (1) the insect fauna; and (2) the islands themselves; and (3) to contribute to a comparative global understanding of oceanic island insect faunas in general.

Foreword

My work on the insects of the Galápagos Islands was inspired and promoted by a series of fortuitous events. As a M.Sc. student at Northwestern University I accidently came upon the synthetic account of island insect colonization and evolution in volume one of Elwood Zimmerman's 1948 classic *The Insects of Hawaii*. As a Ph.D. student at Harvard University I was introduced to field work on island insect biogeography by sampling with Professor E.O. Wilson for 2 weeks on some mangrove islands in the Florida Keys. Then, William A. Shear, a graduate student colleague, showed me a Sierra Club book of photographs of habitats of the Galápagos. I was especially struck by one photo of a humid moss-covered forest in the highlands of Santa Cruz Island. I was sure that such forests had not been properly collected for insects, and vowed to myself that I would do this someday. Another graduate student colleague, Robert Silberglied, soon afterward started a program of insect field work in the Galápagos, but this was abruptly ended when he tragically died in an airplane crash on 13 January, 1982, in Washington, DC. Finally, in 1985, on the occasion of the 150th anniversary of the visit of Charles Darwin to the Galápagos, my wife and I had the opportunity to go to the Galápagos to search for cave and soil insect faunas. Because of the extensive existing literature on Galápagos insects, I originally thought that the insect fauna was relatively well known. I found this not to be true, and took on the expanded research project which led to this volume.

Long ago, Carlquist (1974) wrote of the Galápagos: "Field studies of entomology remain one of the more promising avenues open for investigation". I hope I have been able to show here more fully what lies along this avenue, and to hint at the scientific adventure and discovery that my team has experienced while travelling upon it. But we are not yet at the end of the avenue.

This work is dedicated to all the above people who have inspired me, to all who have contributed to sampling, understanding, and identifying the insects of the Galápagos Islands, and to those whose past and present efforts created and now protect the Galápagos National Park.

Acknowledgements

Research for the 15 months I spent in the field in the Galápagos Islands on five trips was by permit from the Galápagos National Park: F. Cepeda, H. Ochoa Córdova, O. Sarango Valverde, A. Izurietta, and E. Cruz, Superintendents, Department of Forestry, Ministry of Agriculture and Livestock, Republic of Ecuador. Support came partially from operating grants from the Natural Sciences and Engineering Research Council of Canada, and grant 5563-95 of the National Geographic Society. The field work was facilitated by logistical support from the Charles Darwin Research Station (CDRS), Isla Santa Cruz (G. Reck, D. Evans, C. Blanton, Directors). TAME airlines assisted with reduced airfares. My wife, Jarmila Kukalova-Peck, bravely and cheerfully helped on several uncomfortable expeditions and always gave encouragement for the project. She has also helped with illustrations. Field sampling was also helped by J. Cook, J. Heraty, B. Sinclair, B. Landry, C. Vogel, R. Palma, S. Abedrabbo, Elvia Moraima Inca, Maria-Teresa Lasso, T. Finston, L. Roque, and Eduardo Vilema. Charlotte Causton and Lazaro Roque (both of CDRS) provided information on various insects from their own studies. Joyce Cook was of exceptional help in sorting field collections and recording and checking data. Material from museums was loaned by: L. Baert (Royal Institute of Natural Sciences (KBIN), Brussels, Belgium); S. Miller (Bernice P. Bishop Museum (BPBM), Honolulu, Hawaii); D. Kavanaugh (California Academy of Sciences (CAS), San Francisco, California), D. Furth (Museum of Comparative Zoology (MCZ), Harvard University, Cambridge, Massachusetts); and D. Nickle and O. Flint (National Musuem of Natural History, Smithsonian Institution (USNM), Washington, DC. P. Wigfull, Steve Goodacre, Lance Bishop, and Joyce Cook aided with data recording and analysis. Tom Cook and Lance Bishop constructed the regression analyses. The ultimate foundation of accuracy of a study such as this is the many taxonomists who were willing to help with identifications, and they are most gratefully acknowledged individually in the following appropriate chapters.

This is contribution 539 of the Charles Darwin Research Foundation.

Chapter 1
Introduction

The *de novo* appearance of a new volcanic island above the surface of the sea initiates a sequence of physical and biological processes. These start with physical–chemical changes to its originally barren and sterile lava landscapes, and pass through a series of biological events, from arrival of potential colonists, to and through their progressive patterns of establishment, as various habitats and ecological opportunities become available. Just what arrives is a function of dispersal and survival capacity of individual organisms. Successful establishment is a function of the ecology of each species: its needs for habitat and food, and its ecological versatility. This process continues for hundreds and thousands and millions of years after the origin of the island's land surface. Climatic, geographic, and other short- or long-term conditions and events have a progressive and structuring influence on the arriving and evolving biota.

Volcanic archipelagos are among the world's great natural laboratories of evolution. Many studies on the Galápagos, Hawaii, Canary, and other oceanic island groups have shown this repeatedly. They yield special insights on fundamental evolutionary and ecological processes, especially over the first one or two myr (million years) of an island's existance. An exploration of these general patterns and processes, through documentation and analysis of the insects of the Galápagos Islands, especially the the smaller insect orders, is the subject of this book.

The Islands themselves

The Galápagos Islands of Ecuador are a large and complex archipelago. They are famous among biologists and naturalists for the contributions they have yielded to an understanding of the dynamics and mechanisms of distribution and differentiation of biotas. This understanding started with a visit to the islands by young Charles Darwin in September and October 1835. Most of the study of their biota has since focused on the marine life, seed plants, and vertebrate animals. The islands are now one of the most intensively studied areas in the Neotropical Realm, as is evidenced by a very abundant older literature, several large recent volumes of research reports and summaries (Berry 1984; Bowman 1966; Bowman et al. 1983; Grant 1986; Grant and Grant 1989; James 1991; and Perry 1984), and general discussions (Carlquist 1965, 1974; Jackson 1985; Thornton 1971; and Wiggins and Porter 1971). The bibliography of Galápagos references totals over 7500 reports and publications up to 1995 (Snell et al. 1996*b*).

Fig. 1.1. The principal islands of the Galápagos Archipelago, with island contours (at 500 m intervals). Note that Isabela Island was formed by the sequential coalescence of six originally separate volcanos.

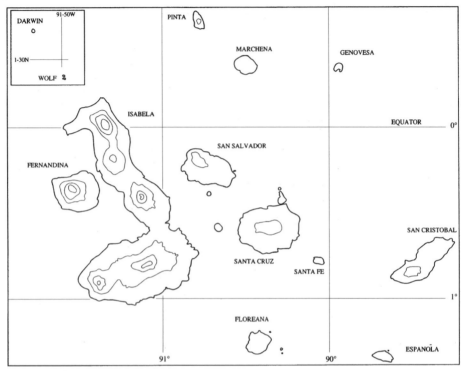

The Galápagos Archipelago contains 127 islands (Fig. 1.1). It lies astride the equator, roughly 1000 km west of the Ecuadorian seaport of Guayaquil, and spans 304 km east to west, and 341 km northwest to southeast. There are 19 large islands over 1 km^2 in area, and some 108 other smaller vegetated islets, and additional wave-washed rocks (Table 1.1). The total land area is approximately 7856 km^2. The largest island, Isabela (4588 km^2), is larger than the total area of all the other islands combined, and is over 4 times as large as the second largest, Isla Santa Cruz (986 km^2). The next six large islands, in order of size, are Fernandina (642 km^2), Santiago (585 km^2), San Cristóbal (528 km^2), Floreana (172 km^2), and Marchena (130 km^2) (Snell et al. 1995, 1996a). The islands are almost entirely volcanic in origin and composition. The exceptions are small areas of uplifted marine sediments deposited on older submarine volcanic substrates (these are on northeastern Santa Cruz, Baltra, Plazas, western Isabela, and North Seymour Islands). Thirteen of the volcanos have erupted in the Holocene and are considered to be active volcanos, and nine of these have erupted in historic times (White et al. 1993). The Galápagos are one of the very few Pacific island groups that was not settled by aboriginal humans before their discovery by Europeans. The Galápagos were discovered by the Spanish in 1535.

Table 1.1. Preferred Ecuadorian and alternate English names of all islands of the Galápagos Archipelago which are known to have insects, and island area and maximum elevation (from Wiggins and Porter 1971, and Snell et al. 1995, 1996a). There is a total of 127 named islands, islets and rocks with terrestrial vegetation, but most of the smaller ones have not been adequately sampled for insects. The focus of this study has been on the large and middle-sized islands. The last column is the total for the island of all insect species in the smaller orders only.

Preferred Ecuadorian Name	Alternate English Name	Area (km²)	Elev. (m)	Small order species
Isla Baltra	South Seymour Island	26.19	20?	27
Isla Bartolomé	Bartholomew Island	1.24	109	14
Roca Beagle Sur	Beagle Rock	0.08	10?	2
Isla Campéon	Champion Island	0.095	46	8
Isla Caamaño	—	0.045	2	2
Isla Cowley	Cowley Island	0.035	30?	2
Isla Daphne Major	Daphne Major	0.330	35?	5
Isla Darwin	Culpepper Island	1.063	168	12
Isla Eden	Eden Island	0.023	30	4
Isla Española	Hood Island	60.48	198	46
Isla Fernandina	Narborough Island	642.48	1494	106
Isla Floreana (Santa Maria)	Charles Island	172.53	640	142
Isla Gardner at Española	Gardner at Hood	0.580	49	7
Isla Gardner at Floreana	Gardner at Charles	0.812	227	13
Isla Genovesa	Tower Island	14.10	76	38
Islote Guy Fawkes	Guy Fawkles Island	0.034	17	1
Isla Isabela	Albemarle Island	4588	1707	211
Isla Marchena	Bindloe Island	129.96	343	49
Isla Pinta	Abingdon Island	59.40	777	76
Isla Pinzón	Duncan Island	18.15	458	60
Isla Plazas Sur	South Plazas Island	0.119	15	12
Islotes Quatro Hermanos	Crossman Islands	0.729	168	1
Isla Rábida	Jervis Island	4.993	367	45
Isla San Cristóbal	Chatham Island	528.09	715	160
Isla Santa Cruz	Indefatigable Island	985.55	864	333
Isla Santa Fé	Barrington Island	24.13	259	21
Isla Santiago (San Salvador)	James Island	584.65	905	146
Isla Seymour	North Seymour Island	1.838	15	28
Isla Tortuga	Brattle Island	1.298	186	8
Isla Wolf	Wenman Island	1.344	253	20

Temporary, but frequent, visits were made by pirate and whaling ships. Permanent human settlement began in 1832, when Ecuador claimed possession of the Archipelago. Six of the islands are now inhabited by nearly 16 000 permanent human residents. The growth rate of the human population is presently 6–7% per year. Some 96.4% of the area of the islands is now included within the protected boundaries of the Galápagos National Park. The remaining land area is composed of small port villages, military bases, three airports, and agricultural areas on the four large islands of Santa Cruz (plus Baltra), Floreana, San Cristóbal, and Isabela. Tourism now brings about 65 000 visitors to the islands each year. The Charles Darwin Research Station, on Isla Santa Cruz, facilitates scientific research and aids the Galápagos National Park Service in training, conservation and management programs.

The Galápagos are of great interest for a variety of reasons. They are truly volcanic, are well isolated (900 km from the nearest continental land-mass of the Santa Elena Peninsula of Ecuador), and are of known age. The older islands are about 3–4 myr in age, the middle-aged islands about 1.5 myr in age and the younger less than 0.7 myr in age. This age spectrum places them, and their biota, in the 'middle age' category of the life-cycle of the world's oceanic volcanic islands and their resultant evolutionary patterns. Each island itself is a semi-replicated and semi-isolated experiment in the processes that originate and shape the patterns of island biology.

There is not enough space here to review the entire natural history of the islands in general. This has, of course, been done in detail for the geology, birds, reptiles, and plants of the Galápagos (Berry 1984; Bowman et al. 1983; Grant 1986; Jackson 1985; Peck 1991; Perry 1984, and references in these works). Similarly, space constraints also necessitate that only some of the voluminous literature can be cited. I have tried to give the most recent citations that will lead into earlier works.

The main purpose of the present chapter is to summarize only the important data which are relevant to the insect fauna of the Galápagos Islands: the world's largest, most complex, and most diverse oceanic archipelago remaining in a still largely pristine condition.

Origin and age

In the drama of Galápagos evolution the archipelago itself forms the theatre. It lies on a submarine platform on the Nazca seafloor crustal plate. Most analyses (Cox 1983) show this plate to be presently moving at a rate of about 3.4–7 cm a year to the east and south towards (and being subducted under) the western margin of the South American crustal plate. A more recent interpretation (Mann

1995) has the Nazca plate moving to the northeast at 5.4 cm a year and the adjoining Cocos plate moving to the northeast at 11.9 cm a year. The islands themselves are volcano "tips," thought to have formed on the seafloor by a "plume" or "hot spot" rising from the earth's mantle. The hotspot itself may have originated some 80 or 90 myr ago, and some rock from this time may now form part of the floor of the Caribbean Sea (Duncan and Hargraves 1984). Some igneous terranes in Costa Rica are thought to have originated at the Galápagos hotspot and moved to the northeast on the Cocos plate, and one (the Quepos terrane) represents the oldest known drowned seamount/oceanic island (Hauff et al. 1997).

The Galápagos are neither especially young nor old islands. Terrestrial colonization has occurred over the 3–4 myr of the existence of the present islands. Physiographic characteristics of the islands are in Tables 1.1 and 1.2. Radiometric dating of isotopes in the lavas shows that the older southeastern islands of Española and San Cristóbal may have been available for terrestrial colonization from about three to four myr ago up to the present (Cox 1983; Geist 1996). The central islands of Santa Cruz, Floreana and others became available from about 0.7–1.5 myr ago, and Isabela, Santiago, Fernandina (and the other smaller islands to the north) appeared less than 0.7 myr ago. Santiago, Pinta, Isabela, Fernandina, and Marchena islands have experienced very extensive recent volcanic activity and have large or very large surface areas of young and barren lava.

There have been earlier suggestions that the Carnegie and Cocos submarine ridges, to the east and northeast of the Galápagos, once provided continuous dryland corridors for the biotic colonization of the Galápagos. Another idea is that the Galápagos are a fragment of the South American plate which has separated from it and drifted to the west (Rosen 1975). But present understanding of sea floor plates and their direction of movement excludes either idea. There is now no serious evidence to suggest that the islands ever had direct land–bridge connections with the Central or South American mainland. They are true oceanic islands, have always been surrounded by large oceanic spaces, and all terrestrial colonists have had to cross the oceanic water barrier to reach them. The biota is, therefore, entirely assembled by dispersal processes, and there is no reason to believe that there is any old plate tectonic or vicariant component in the biota.

However, there is the recent discovery of drowned seamounts (guyots), with beach-worn cobbles on their summits, on the Carnegie Ridge east of the Galápagos. This shows that earlier volcanic islands may have existed 5–9 myr ago or even earlier and these may possibly have served as stepping-stones for colonization by ancestors of some of the present terrestrial fauna (Christie et al. 1992). Even older islands were on the Cocos Ridge, northeast of the Galápagos, and they are now incorporated into mainland Costa Rica as the Quepos terrane

Table 1.2. Additional characteristics of the principal islands of the Galápagos Archipelago (ranged from SE to NW), and volcanos of Isabela Island. Ages are from Geist (1996). Ecological complexity is measured as number of major vegetation zones. Insect species numbers are for all the small orders and only the native species (endemic plus indigenous species). There is generally a positive correlation in total numbers of insect species in the smaller orders with island area, elevation, and ecological complexity, but not age.

Island	Approximate age (myr)		Ecological complexity	Total of all small order species	Total of native species only
	min.	max.			
San Cristóbal	2.3	6.3	6	160	116
Española	2.8	5.6	2	46	36
Santa Fé	2.8	4.6	2	21	17
Florena	1.5	3.3	4	142	113
Santa Cruz	2.2	3.6	6	333	240
Rábida	1.0	2.5	3	45	37
Pinzón	1.4	2.7	4	60	52
Isabela	0.06	0.70	6	211	155
Volcan Wolf	0.06	0.30			
Volcan Cerro Azul	0.06	0.30			
Volcan Sierra Negra	0.07	0.70			
Volcan Darwin	0.07	0.70			
Volcan Alcedo	0.15	0.30			
Santiago	0.77	2.4	6	146	107
Genovesa	<0.7		2	38	28
Marchena	<0.7		2	49	39
Pinta	<0.7		4	76	61
Fernandina	0.006	0.30	4	106	82

(Hauf et al. 1997). Biochemical (allozyme) dating of the time of origin of the two lineages of endemic genera of Galápagos iguanas can be interpreted as support for the sunken stepping-stone colonization model (see Wyles and Sarich 1983). So also can allozyme data on a cladogenetic split of 6 myr in *Stomion* darkling beetles (Finston and Peck 1997). Thus, terrestrial evolution may actually have been occurring in the vicinity of the archipelago for much longer than the ages of the present islands themselves.

It must be remembered that the island of Isabela is now a composite resulting from the coalesence of four northern volcanos (Ecuador, Wolf, Darwin, and Alcedo) and two southern ones (Sierra Negra and Cerro Azul). All originated as separate individual islands, and the northern and southern groups were only very recently joined at Isabela's low elevation "waist" by the unvegetated recent lava flows of the Perry Isthmus between Sierra Negra and Volcan Alcedo.

Fig. 1.2. Comparative larger areas of the islands at lower sea levels (130 m lower than present) during the last glacial maximum. The dark areas represent additional island surfaces exposed above sea-level during the last glacial maximum, and grey regions are the present islands (after Geist 1996).

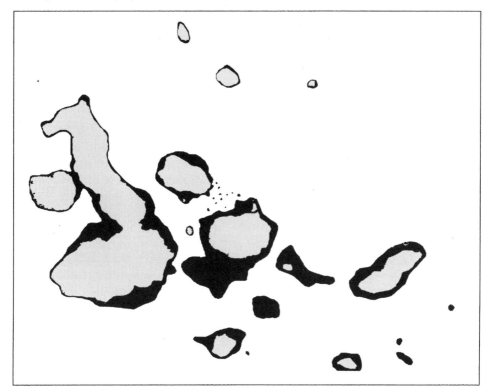

The areas of the volcanically active islands increased as they grew through the addition of new lavas. Conversely, areas of the islands with dead volcanos decreased through erosion and island subsidence. Sea levels themselves have risen and fallen in the Pleistocene, and this too has altered the areas of the islands. During the last glacial maximum (at 17 000 years ago) the sea may have been as much as 130 m lower than now and the island areas were even larger than at present (Fig. 1.2). But the main islands still remained separated at these times of low sea levels because the channels between these islands are deeper than 130 m.

Present currents and climate

The climate of the islands is arid to semi-arid in their lower elevations. They lie at the edge of the weather region called the Central Pacific dry zone. The higher islands have increasingly more rain at progressively higher elevations

7

and a humid forest zone exists at 300–600 m on their windward sides. Above this lies a climatic inversion zone and progressively drier conditions from 1000 to 1700 m. Precipitation is variable in occurrence and quantity, even in the wet season. Total annual rainfall in coastal semi-arid areas may be as much as 75 cm. During the cool/dry or "garua" season, from June to December, the daily temperature range is 19–23°C. At this time, there is prolonged cloud cover or overcast, and little or no rain in the lowlands. Also, at this time, all areas from about 400 to 1000 m elevation are usually embedded in perpetual drizzle and clouds. January to May is the warm/wet season, usually with sunny skys, occasional rain, and with a daily temperature range of 24–29°C (see Colinvaux 1984, Grant 1986, or Jackson 1985 for more details).

The climate of the islands is controlled by both the winds and oceanic currents which bathe them. Both these come predominantly from the northeast, east, or southeast and their direction changes in response to a seasonal north-south shift in the Intertropical Convergence Zone (ITCZ). The warm-wet (El Niño) season (more technically an El Niño-Southern Oscillation or ENSO event) occurs when the ITCZ has moved to the south to lie near, over, or south of the islands. It brings warm and moist air and warm oceanic currents from the region of the Gulf of Panama (Fig. 1.3). At intervals of 3–6 (7 ± 4) years this change becomes very pronounced, a strong El Niño occurs, and extraordinarilly heavy rains fall, which may be 10 times the normal annual amounts. In contrast, in the cool-dry season, the ITCZ moves much to the north of the Galápagos and the islands are influenced by cool-dry winds and cool ocean currents (the Peru or "Humboldt Current") from the arid coast of South America. These cool currents are responsible for both the pleasant-to-cool nighttime temperatures and the high nighttime relative humidity; which can be up to 90%, even in the dry season (Fig. 1.4). These nighttime climatic conditions make possible much of the present diversity of terrestrial arthropods. It seems an unappreciated fact that the high relative humidities, dew, and garua mists, all of which occur at night-time in the arid zone, may be the most important factor allowing the insect fauna to be as rich as it is. Insect activity and abundance is ever so much greater at nighttime than in the heat, low humidity, and bright sun of the day. For example, the number of individuals of "Darwin's darklings" (flightless tenebrionid beetles) active at night is positively correlated with nighttime relative humidity (Finston et al. 1997).

Past currents, geography, and climate

I suggest that the major present-day patterns of east equatorial Pacific wind and oceanic currents and the climates that they create may have become established at about the time of the first appearance of the present islands above the sea, about 3–4 myr. This would roughly coincide with the closure of the Tertiary seaway across the Isthmus of Panama (Jackson et al. 1996) and development of

Fig. 1.3. Simplified diagram of seasonality and source of major wind and oceanic currents influencing the Galápagos Archipelago (G), and Cocos (C) and Malpelo (M) islands in January (from Graham 1975). When the cool westward flowing South Equatorial Current (SEC), composed of a Peru coastal and a Peru oceanic current, bathes the Galápagos, from about June to November, the islands generally have a cool and dry climate in the lowlands and the garua mists of the highlands. From December to May the SEC and Intertropical Convergence Zone are closer to or south of the Galápagos and a warmer oceanic El Niño flow with wetter air masses comes from the Panama Basin. The Equatorial Countercurrent is usually north of the Archipelago, but can move south of it, and is most strongly developed in August and September. Normal El Niño conditions are strengthened at about 3–6 year intervals when a stronger El Niño flow comes from the Panama Basin and surrounds the Galápagos with even warmer but nutrient poor waters and even more warm-moist air. These changes are initiated by poorly understood conditions in the west Pacific. The El Niño winds and currents have carried most of the terrestrial invertebrate colonists from Pacific coastal Mexico, and Central and northern South America.

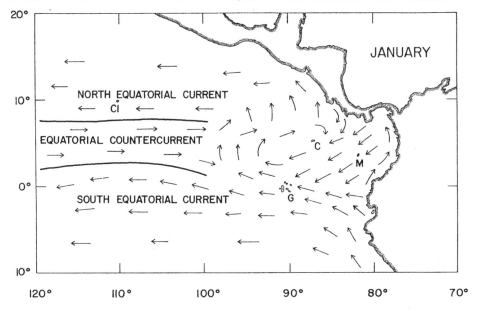

the present major patterns of movment of east Pacific Basin winds and ocean currents.

Pleistocene climatic changes have certainly had their impact on the islands. But the present large islands have always remained separated from each other, even while sea levels were lower by 100 m or more during full-glacial episodes (Fig. 1.2). The sea channels between the major islands are all deeper than this. The major difference is that many of the large islands were thus somewhat closer to each other than they are now (Geist 1996).

A limited palynological record from El Junco Lake on Isla San Cristóbal shows that the islands have not been appreciably wetter than they are today. In the past glacial, and probably earlier glacials, they were actually markedly drier

9

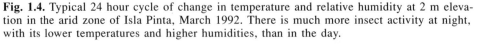

Fig. 1.4. Typical 24 hour cycle of change in temperature and relative humidity at 2 m elevation in the arid zone of Isla Pinta, March 1992. There is much more insect activity at night, with its lower temperatures and higher humidities, than in the day.

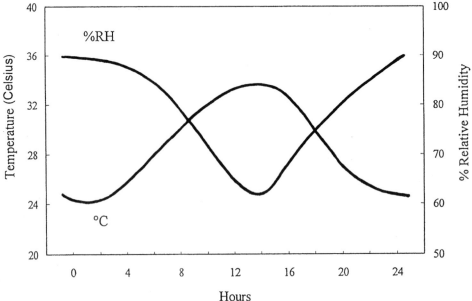

than they are now (i.e., they were without the present significant rainy season). Much of the vegetated area of the wetter upland forests was diminished or even absent (Colinvaux 1972, 1984; Colinvaux and Schofield 1976).

There is evidence from coastal northern Peru which is interpreted as showing that El Niño and ENSO climatic events only began about 5000 years ago (Sandweiss et al. 1996). If this is also true for the Galápagos, it would alter some or all of the above views of the early and mid-Holocene climate of the islands. But there is disagreement about the validity of the Peru evidence (DeVries et al. 1997).

Biotic or vegetational life zones

The composition and structure of the vegetation of the islands changes with distance from the seacoast and increase in elevation. Six major natural vegetational zones are generally recognized. Additional subdivisions could also be recognized. On the windward sides of the islands, the vegetational zonation changes from (1) a salt-tolerant littoral zone, to (2) a lowland arid zone of cactus and microphyllous thorn forest, (3) a mixed transition forest zone, (4) an evergreen macrophylous humid forest (or *Scalesia*) zone, (5) a sub-treeline evergreen shrub (or *Miconia*) zone, and (6) a fern-sedge "pampa" zone which lies above regional treeline (Fig. 1.5). On the highest points of the two highest

Fig. 1.5. Major vegetational zones of Santa Cruz Island. Zonation is controlled by elevation and moisture, which is related to the prevailing winds. Low islands have only a littoral and arid zone. Some high islands lack the zones characterized by *Scalesia* and *Miconia*, but have equivalent humid forest and shrub zones dominated by other woody genera (vertical scale greatly exaggerated).

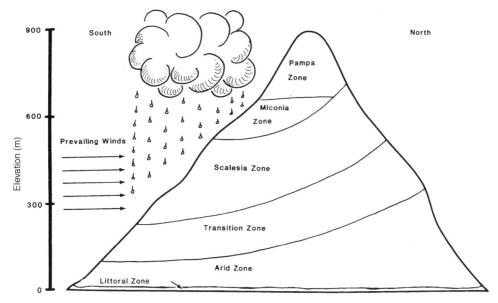

volcanos on Isabela lies (7) an upper arid zone, above the climatic inversion layer. An additional eighth zone can be recognized as the one created by the settlement and agricultural activities of the human population.

On the leeward side of high islands the arid and transition zones rise higher and a humid forest zone may even be absent (Wiggins and Porter 1971). Treeline for indigenous and endemic woody vegetation is at about 700 m. That is, the locally-evolved woody vegetation has been unable to adapt to climatic-physiological conditions above this elevation. This is in contrast to a treeline at about 3500 m in the Andes of mainland Ecuador, at the same latitude. Adventive trees, such as cinchona and guava, however, are able to live above the islands' locally evolved treeline, and are a serious ecological threat to the endemic plant communities of the naturally treeless highlands.

Several authors have described the vegetational zones, especially on Santa Cruz (Reeder and Reichardt 1975; Wiggins and Porter 1971), and the zones are generally similar in their structure the other large "high" islands of San Cristóbal, Floreana, Santiago, Isabela, and Fernandina. The zones can be characterized in more detail as follows.

1. The littoral zone. This is a narrow coastal belt, extending from somewhat

below the high tide line and inland from 10 to 100 m or more as a zone of salt-tolerant vegetation of halophilic mangroves, succulents, and salt marsh.

2. The arid zone. This zone may comprise the entirety of the remaining area of the small or low islands. This is a microphyllous xerophytic vegetation and extends up to 80–120 m on the southerly faces of the islands, and up to 200 or 300 m or more on the leeward side of the larger islands. The dominants in the vegetation are arborescent *Jasminocereus* and *Opuntia* cacti, and seasonally deciduous trees in the genera *Prosopis*, *Acacia*, and *Bursera*.

3. The transition zone. This zone receives more moisture, and some of the plants are evergreen. The trees are taller and more numerous than in the arid zone. Tree species of *Pisonia*, *Psidium*, and *Piscidia* are dominant. The plant species here also occur in either the arid or humid forest zones.

4. The humid forest (or "*Scalesia*") zone. This zone contains a meso-phyllous, mainly evergreen forest, and is characterized on Santa Cruz, Santiago, San Cristobal, and Floreana by the endemic tree composite *Scalesia pedunculata*. Another arborescent *Scalesia* species occurs on Cerro Azul and Sierrra Negra on southern Isabela and a third on Fernandina and volcans Wolf, Darwin, and Alcedo on northern Isabela (Itow 1995). In addition, other trees such as *Zanthoxylum fagara* and other genera also occur in the humid forest. This zone has been extensively altered by clearing for agriculture on Santa Cruz, Floreana, and southeastern Isabela and by overgrazing by goats on San Cristóbal. The zone's elevational range may be from 180 to 550 m. Rainfall (at Bellavista, Santa Cruz) averages 1040 mm. When well developed, the forest occurs with a rich and thick undergrowth, made possible by the increased moisture levels (Eliasson 1984). Subdivisions of this zone could be closed-canopy evergreen forest and mossy evergreen forest. A few small disjunct patches of this forest reappear where mountain crests catch condensing water vapor, such as the windward crests of Volcan Fernandina, Volcans Alcedo, Darwin and Wolf on Isabela Island, and the highest point on Floreana Island.

5. The evergreen shrub (or "*Miconia*") zone. This zone occurs above the humid forest zone. On Santa Cruz, Santiago, and San Cristóbal it is dominated by the endemic melastome shrub *Miconia robinsoniana* from about 450 to 625 m. This plant may form almost pure stands of virtually impenetrable 3–4 m tall mesophyllous evergreen shrubs with an almost closed canopy, and stems and branches which are heavily sheathed in epiphytic moss, lichens, and liverworts. Rainfall (at Media Luna, Santa Cruz) averages 1694 mm. The zone is roughly equivalent to the montane subparamo (elfin forest) zone of the Andes of mainland Ecuador.

6. A fern-sedge or "pampa" zone. This replaces the evergreen shrubs at elevations of 550–650 m and extends to the tops of the mountains on some of the

high islands, such as Santa Cruz and San Cristóbal, and forms large areas on Volcan Cerro Azul and Sierra Negra on Isabela. These last four areas may be larger then they were originally because of recurrent fires and prolonged grazing by feral cattle. Indigenous woody vegetation is absent. A distinct pampa zone is not well developed on the high island of Fernandina, and on the volcanos of Darwin and Wolf on Isabela Island. The summit of Pinta is mostly covered

Plate I. A–F. Habitats of the Galápagos islands. (A) Isla Santa Cruz from littoral to pampa zone: six major vegetation zones in an elevation rise of 700 m. (B) Succulent *Sesuvium* and *Heliotropium* vegetation of the littoral zone of Isla San Cristóbal. (C) A collecting team crossing young and barren lava flow on Isla Fernandina. Left to right, Eduardo Villema, Claus Vogel, John Heraty. (D) Evergreen shrub zone on Isla Santiago. (E) Moss and lichens in an invasive guava grove in the humid forest zone of Isla Santa Cruz. (F) Permanent freshwater lake in crater at summit of Isla San Cristóbal.

with humid forest and bracken fern. The summit areas of Santiago and Isabela's Volcan Alcedo have been degraded from an open canopy humid forest to a pampa through the grazing activities of feral goats (Hamann 1993; Desender et al. 1999).

7. Upper arid zone. The two tallest volcanos (Cerro Azul and Volcan Wolf) on Isla Isabela penetrate the upper inversion layer and have extensive high elevation areas of typical arid zone vegetation, with arborescent *Opuntia* cacti or *Scalesia* as the dominant plants. The plant and animal species of the upper arid zone all seem to be ones also found in the low elevation arid zone of the island. There are no known insect species unique to this upper arid zone.

All of the above vegetation zones merge one into another (Eliasson 1984; Reeder and Reichardt 1975) and sharply defined ecotones do not normally exist except at the edge of the evergreen shrub and pampa zones, and this may have been caused by past fires, as on Santa Cruz. The effects of past fires on the extent of the life zones have not been studied.

8. The urban and agriculture zones. The vegetation of some of the arid zone, transition zone, and especially the humid forest and evergreen shrub zones has been extensively altered by clearing for villages, farms, crop agriculture, and pastures on Floreana, south-eastern Isabela, Santa Cruz, and much of the highlands of San Cristóbal. This has created human-influenced urban and agriculture zones, and these are usually dominated by adventive plants for agricultural or horticultural purposes or by weeds.

The insects

The first insects were collected on the islands by Charles Darwin during his 1835 visit. He was there during the dry season in September and October, was on land for 19 days, and only on the islands of Santiago, Isabela, San Cristóbal, and Floreana (Smith 1987; Sulloway 1984). He found the insect fauna to be very poor and of little interest. His summary of his experience is quoted in the prologue of this book. Since then, many scientists have added to knowledge of the insect fauna (Fig. 1.6). Slevin (1959) provides a history of all the early explorations. Scientifically, the most significant of these was the year-long pioneering 1905–1906 California Academy of Science (CAS) expedition with F.X. Williams as entomologist. The next most significant expedition was that of the Galápagos International Scientific Project (GISP) of 1964 organized by the University of California and with six entomologists (E.G. Linsley, R. Usinger, G. Kuschel, D. Cavagnaro, P.D. Ashlock, and R.O. Schuster). Usinger (1972) briefly describes this expedition. Linsley and Usinger (1966) summarize the 22 earlier expeditions which made entomological collections. These and some by a few other workers brought the known insect species total of the islands to 883 by 1977

Fig. 1.6. Year and duration of visits of historical insect collecting expeditions to the Galápagos (some data from Linsley and Usinger 1966). This is no more than a rough generalization of historical insect sampling effort, because records do not usually exist to determine the number of people involved or the amount of effort spent collecting. For instance, Charles Darwin was in the Archipelago for 37 days, but was on land for only 19 days (and some only in part), and the 1923 "Harrison Williams" expedition of William Beebe was in the islands for 21 days, but only 100 h were spent on shore (Slevin 1959). Of course, there have been many other scientific expeditions (see Slevin 1959), but they made no or few entomological contributions. Most of the sampling has been done in the favorable rainy season, from January to June.

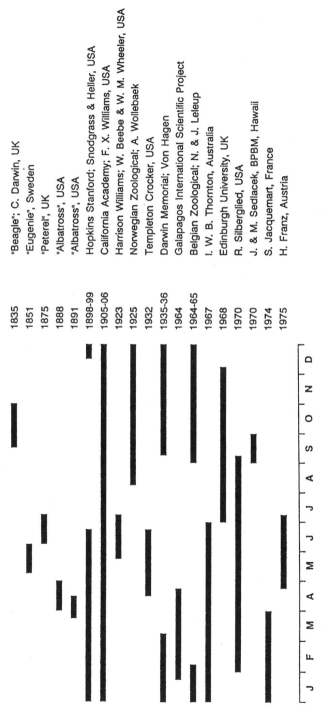

SEASONAL ACTIVITY OF IMPORTANT INDIVIDUALS OR EXPEDITIONS COLLECTING INSECTS IN THE GALAPAGOS

1835	"Beagle"; C. Darwin, UK
1851	"Eugenie", Sweden
1875	"Petere!", UK
1888	"Albatross", USA
1891	"Albatross", USA
1898-99	Hopkins Stanford; Snodgrass & Heller, USA
1905-06	California Academy; F. X. Williams, USA
1923	Harrison Williams; W. Beebe & W. M. Wheeler, USA
1925	Norwegian Zoological; A. Wollebaek
1932	Templeton Crocker, USA
1935-36	Darwin Memorial; Von Hagen
1964	Galapagos International Scientific Project
1964-65	Belgian Zoological; N. & J. Leleup
1967	I. W. B. Thornton, Australia
1968	Edinburgh University, UK
1970	R. Silberglied, USA
1970	J. & M. Sedlacek, BPBM, Hawaii
1974	S. Jacquemart, France
1975	H. Franz, Austria

J F M A M J J A S O N D

15

Table 1.3. Summary of the growth of knowledge of the insect fauna of the Galápagos Archipelago, Ecuador.

All Insects				
Year	Orders	Families	Genera	Species
1835–1966[1]	19	129	395	618
1966–1977[2]	22	164	531	883
1977–2000[3]	24	255	1057	1850+

[1]data from Linsley and Usinger 1966;
[2]data from Linsley 1977;
[3]data from above plus new literature and results reported here.

(Table 1.3), as documented in a total of 308 scientific publications on the insects alone. Additionally, invaluable new data on the spider fauna (Baert et al. 1989*a* and *b*) and mites (Schatz 1991, 1998) has provided an improved listing for these important groups of terrestrial arthropods (Table 1.4). When the published data are added to the results of my five expeditions, I estimate that 90–95% of the total species diversity of the insect and terrestrial arthropod faunas of the Galápagos is now sorted or sampled, but it is not yet all named to species or published (Table 1.5).

How many insect species actually occur in the Galápagos?

Of interest in such a study is the question "How many insect species actually occur in the Galápagos?" We do not yet know all of them. My direct-count approach to documenting species abundance has proved to be very time-consuming and laborious. At the start of this study I attempted two indirect estimates of the total number of the insect species which might be present.

1) *The insect–host plant association method.* One method is based on inferences of the number of insects associated with their host plants. These numbers may be highly variable, and they depend on the geographic area and abundance of the host plant, the evolutionary time to originate such associations, and the size (structural complexity) of the host. Trees support more insect species (because they offer more "niches") than do shrubs, herbs, monocots, or ferns. A plant species in Britain frequently is host to more than 50 insect species (Strong and Levin 1979; Lawton and Schroder 1977). Few data are available for tropical islands. Southwood (1960) summarized data for trees on Hawaii and found from 1.8 to 41 species of insects (averaging 40 species) obligately restricted to a single tree species and a total of from 5.6 to 155 insect species (averaging 123 species) associated in some way with each tree species. On the Mediterranean island of Cyprus, Southwood (1961) reports 3.4 species of Heteroptera and Auchenorrhyncha alone per tree species. These figures may be too high to apply directly to the Galápagos. Cyprus is an island considerably larger and nearer to surrounding continental

Table 1.4. Known diversity of species of Onychophora and terrestrial and freshwater Arthropoda (excluding Hexapoda) of the Galápagos (from Baert et al. 1989*a*, *b*, 1995; Baert 1994, and in prep.; Gertsch and Peck 1992; Peck 1994; Schatz 1991, 1998; Shear and Peck 1987, 1992; Peck and Shear 2000, 2001). Numbers will change as the faunal studies continue, especially in the Acari (mites) where the total species number is expected to be 360–460 species (200–320 endemic) (Schatz, unpublished).

				Species thought to be:		
	Families	Genera	Species	Endemic	Indigenous	Adventive
Onychophora	1	1	1	0	0	1
Arthropoda						
Scorpiones	2	2	2	2	0	0
Amblypygi	1	1	1	1?	0	0
Schizomida	1	1	1	0	0	1
Araneae	28+	61+	156+	62+	90+	2+
Solifugae	1	1	1	1	0	0
Pseudoscorpiones	4?	13	18	13	5?	0
Opiliones	1	1	3	3	0	0
Acari						
Gamasina		38	30	8	0	
Uropodina		13	11	2	0	
Ixodina		14	10	4	0	
Actinedida		52	46	6	0	
Acaridida		1	0	1	0	
Oribatida	64	125	202	81	112?	9?
Crustacea						
Notostraca	1	1	1	0	1	0
Conchostraca	1	1	1	0	1	0
Anostraca	1	1	2	1	1	0
Ostracoda	2	3	3	1	2	0
Cyclopoida	5	7	7	0	7	0
Decapoda	2	2	3	0	3	0
Amphipoda	1	3	4	1	3	0
Isopoda	10	13	17	4	6	7
Chilopoda	6	10	13	6	2	5
Diplopoda	7	9	10	1	0	9
Symphyla	1	1	1	0	0	1

source areas. Hawaii is exceptionally remote, but much larger and older than the Galápagos, and more in situ speciation there has enriched both the flora and the

Smaller Orders of Insects of the Galápagos Islands: Evolution, Ecology, and Diversity

Table 1.5. Summary of insect faunal diversity known from the Galápagos Archipelago, and estimates of minimum number of natural colonizations needed to originate the endemic and indigenous species. Numbers will change somewhat as the faunal study continues. New specimens column gives totals of new specimens segregated to species from samples of my collecting teams and in the collections of BPBM, CAS, CDRS, KBIN, MCZ, USNM, etc., These, plus the specimens reported in the existing literature, are all used as the basis for the conclusions presented here.

| Orders | Families | Genera | Species | Number of species: | | | Natural Colonizations | New Specimens |
				Adventive	Indigenous	Endemic		
Collembola	7	24	35	3	22	10	33	200
Diplura	2	2	2	1	1	0	1	15
Archeognatha	1	1	1	0	0	1	1	226
Thysanura	2	3	3	1	1	1	2	455
Odonata	3	7	8	0	7	1	8	282
Orthoptera	3	16	33	4	1	28	16	5 705
Mantodea	1	1	1	0	0	1	1	85
Blattodea	4	12	18	11	2	5	3–4	2 606
Isoptera	2	3	4	0	3	1	4	1 416
Dermaptera	2	5	7	4	1	2	2	1 424
Embidina	2	2	2	1	0	1	1	39
Zoraptera	1	1	1	0	1	0	1	14
Psocoptera	15	22	39	13	6	20	20	0
Thysanoptera	4	31	50	8	42	0	49	1 435
Phthiraptera	5	40	80	1	66 ?	13 ?	80	3 000
Hemiptera	20	71	129	16	35	78	83	6 697
Homoptera	17	70	141	56	13	72	50 ?	14 547
Neuroptera	3	5	8	0	3	5	8	1 282
Strepsiptera	1	1	1	0	1	0	1	27
Coleoptera	59	238	420+	61	101	258	313 ?	—

Table 1.5 (*concluded*).

Orders	Families	Genera	Species	Adventive	Number of species: Indigenous	Endemic	Natural Colonizations	New Specimens
Lepidoptera	22	168	346+	46	±200	±100	250+	—
Siphonaptera	3	4	4	3	0	1	1	5
Diptera	47	161	240+	39	93	108	200 ?	—
Hymenoptera	29	169	280+	31	220+	29+	249	—
Totals	255	1057	1853+	295+	818+	736+	1375+	40 002

Table 1.6. Characteristics of some oceanic islands and their insect faunas. Beetles are given in detail because this is the largest order of animals, and are often the best known part of an insect fauna. The Pacific islands are discussed by Gressitt (1961); for the Indian Ocean see Paulian (1961) and Peake (1971).

Island or Archipelago	Present number main islands	Present total area (km^2)	Time available for terrestrial colonization (myr)	Shortest distance to mainland (km)
A. Atlantic Ocean				
Ascension	1	97	1.5	2600
Azores	9	2 300	0.4–8	1400
Bermuda	7	54	0.1	1040
Canary	7	7 502	1–20	90
Cape Verde	9	4 032	?	600
Madeira	2	790	?	650
Selvagens	2	4.5	30	500
St. Helena	1	120	15	1800
B. Indian Ocean				
Aldabra	4	155	0.038	420 (Madag.) 640 (Africa)
Seychelles	30	404	?	1450
C. Pacific Ocean				
Cocos	1	27	?	500
Easter	1	530	2.5–3	3700
Galápagos	9	7 856	0.7–3.7	800
Guam	1	559	?	2400?
Hawaiian	8	16 615	0.5–5	4000
Juan Fernandez	3	85	3	650
Marquesas	10	1 036	Pliocene?	4830
Samoa	4	3 121	?	4000?
Society Islands	14	1 684	?	5000?

plant-feeding fauna. Basset et al. (1996) discuss in greater detail these and other problems in measuring numbers of host specific insects.

Nevertheless, if we take as a conservative figure 4 insect species to be host specific for each of the 560 indigenous species of vascular plants on the Galápagos, there could be a fauna of over 2000 phytophagous insect species. If there is about an equal number of other (predatory and saprophagous) species (Strong et al.1984) the Galápagos total would be some 4000–5000 insect species.

Table 1.6 (*concluded*).

Number insect species	Number beetle endemics	Percent beetle endemics	Principal or most recent references
179	38	0	Ashmole and Ashmole 1997
1531	522	11.3	Wollaston 1857; Méquignon 1946; Serrano 1982; Gillerfors 1986; Borges 1990
?	228	0	Hilburn and Gordon 1989
c.5000	1728	55	Wollaston 1864, 1886; Baez 1984, 1987; Machado and Oromí unpubl.
?	470	31	Uyttenboogaart 1946; Geisthardt 1988
?	c.900	39	Lindberg 1963; Erber and Hinterscher 1988; Lundblad 1958
87	42	50	Oromí 1983
?	256	61	Basilewsky 1985
1000	176	?	Cogan et al. 1971
(39% end.)	?		
2090	?	65	Scott 1933
333	63	?	Hogue and Miller 1981
?	56	0	Desender and Baert 1996
1850	359	72	
?	206	?	Swezey 1942, 1946
7979	1984	68	Zimmerman 1948, 1970; Howarth 1990; Eldredge and Miller 1997
687	235	81	Kuschel 1963; Peña 1987
450	124	?	Adamson 1939
1603	385	52	Buxton 1935
470	?	?	Cheesman 1927, 1934

Complicating these estimates are observations (Becker 1975) that islands support more generalist feeders (scavengers and saprophages) and fewer specialist feeders (phytophages) than do continents. Later study by us showed very little evidence for host specificity for plant-feeding insects and demolished this approach to making an estimate. Studies of obligate and facultative insect associations with single Galápagos plant species are still an obvious research area for Ecuadorian or other students. The only obligate host associations that come to mind for the insects are *Gerstaeckeria* weevils which feed only on *Opuntia* cactus and the Monarch butterfly which feeds only on adventive milkweed plants.

2) *The species–area relationship method.* Another method of estimating the size of the insect fauna is by examination of the relationship between number of species and island area. The argument is based on many observations that species numbers increase in a regular way as island area increases. From such knowledge of general island insect faunas one can sometimes extrapolate to unstudied islands.

Unfortunately, the insect faunas of tropical and subtropical islands are unevenly known. Comparative data on some island insects are in Table 1.6. It turns out that data on beetles are more complete than for other insect groups. The beetle data can be used to calculate a species-area regression as in Fig. 1.7. Note that this questionably combines total numbers from archipelagos and compares

Fig. 1.7. Regression of numbers of tropical oceanic island beetle species against areas of individual islands or archipelagos. Data from Table 1.4. The Canaries and Hawaii figures include adventive species. The Galápagos fall far below the regression line, indicating that they seemingly have far fewer species than is normal for islands of their size. This suggests either that many more species are to be expected (but this is now unlikely) or that their harsh aridity and isolation have really limited the number of species that have been able to naturally colonize and speciate on the islands.

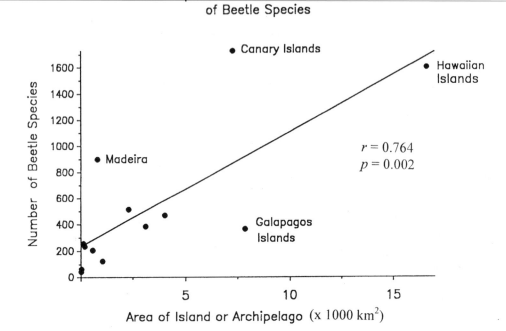

Relationship Between Island Area and Number of Beetle Species

$r = 0.764$
$p = 0.002$

them with numbers from individual islands. If the line is accepted as being a reasonably correct first-order approximation of equilibrium saturation numbers of species, it suggests that the Galápagos might actually have up to about 800 beetle species for its area, all other factors being equal. If the species composition of an insect fauna is 30% beetles (as is usually asssumed for most assemblages of terrestrial animal species, Wilson 1992) this suggests a total insect fauna of about 2650 insect species. In contrast, the results of the present study have found 359 native species of beetles (plus 61 adventives), and a total of somewhat less that 1600 species of native insects. This suggests that equilibrium-saturation numbers for the Galápagos are really lower than for other islands, and this is probably caused by the prolonged seasonal aridity of the islands, which limits the survival and establishment of colonists and their descendants.

Data from more oceanic islands are needed to judge the multiple factors that have an influence on the resultant richness of an island's insect fauna. These factors may be a single island vs. an archipelago, nearness to mainland or other "stepping stone" island source areas, island age, island area, island elevation, climatic conditions such as seasonality of temperature or rainfall, and so on. The data used above also frequently do not separate native from adventive species in the faunal totals. We will probably never know the true original insect richness of islands now greatly changed by humans, such as the Azores, Bermuda, the Canaries, the Cape Verdes, St. Helena, Mauritius, and many others.

3) *A direct count.* The presently known insect diversity is at least 1558 native species (823 indigenous and 735 endemic species; Table 1.5). An additional 292+ species are adventives (accidental introductions by humans in historical times). This count is based on all previous literature records, and all new specimens which have been collected and identified or segregated to species units by myself, my team, and many cooperating taxonomic specialists. For the smaller orders alone this represents a total of 42 002+ specimens additional to those included in all the previous literature. Many more uncounted specimens are involved in achieving the species totals for the 5 orders not covered in detail in this book.

These direct count figures should serve to show the completeness and weight of the new data, and that the two higher indirect estimates are most unlikely to be ultimately supported. I now estimate that the actual total number of native insect species in the Galápagos will not exceed 2000.

Total Galápagos biodiversity

The Galápagos are now one of the most intensively biologically studied sites in the world. It is of some interest to ask how marine and terrestrial species diversity compare and contrast, and how insects compare to other terrestrial groups. Even though the sea represents about 70% of the earth's surface it seems

Table 1.7. Comparative species diversity in marine versus terrestrial (plus freshwater) habitats in the Galápagos. Native species totals are indigenous and endemic species combined.

	Number of native species	Number of endemic species
MARINE HABITATS OF THE GALÁPAGOS		
"Plants"		
Algae	333	116
Vertebrates		
Fishes	447	51
Mammals (cetaceans and pinnipeds)	24	2
Invertebrates		
Meiofauna in marine beach sands (9 phyla)	390	?
Coelenterata, Scleractinian corals	44	20
Coelenterata, Gorgoneans	12	8
Platyhelminthes	95	85?
Mollusca	800	141
Bryozoa	184	34
Annelida, Polycheta	192	50
Arthropoda, Chelicerata	54	44
Crustacea	432	143
Insecta (*Halobates*)	3	0
Echinodermata	200	34
Other miscellaneous invertebrates	70	42
Totals	3 280	770 (23.5%)
TERRESTRIAL AND FRESHWATER HABITATS OF THE GALÁPAGOS		
"Plants"		
Macro-fungi	59	0
Flowering plants	560	180
Mosses and liverworts	204	22
Lichens	213	11
Ferns and allies	108	8
Vertebrates		
Reptiles	20	21
Birds (including 19 sea birds)	67	27 (5)
Mammals	7	9
Invertebrates		
Tardigrada	14	2
Nematoda	100+	5?
Mollusca, land snails	83	80
Spiders	80	55

Table 1.7 (*concluded*).

	Number of native species	Number of endemic species
TERRESTRIAL AND FRESHWATER HABITATS OF THE GALÁPAGOS (contd.)		
Mites	±400	±275
Other Arachnids	27	21
Crustacea	21	5
Centipedes and Millipeds	8	6
Insects	1 555	735
Totals	3 526	1462 (41.5%)

that the greatest biotic diversity is in terrestrial habitats, which are known to contain about 65% of all the world's known species (Wilson 1992). Table 1.7 is a summary of known species diversity by Phylum, Class, or Order, segregated by terrestrial or marine habitat. Certainly, all groups are not adequately known. Some marine invertebrates have not been summarized, such as the non-coral coelenterates, sponges and some crustacea (especially zooplankton), and groups such as rotifers, sipunculids, echiurids, etc. have not been studied. Unstudied terrestrial and freshwater groups such as nematodes, tardigrades, copepods, and oligochaetes will similarly increase the numbers of terrestrial and freshwater species. An early attempt showed Galápagos terrestrial diversity to be a bit richer than that in adjacent oceanic habitats (Peck 1993).

Now, the present total richness of all plants, fungi, and animal species in terrestrial and water-interface habitats stands at 3436 species. This is about 1.07 times larger than the total of 3208 species for marine habitats. Thus, terrestrial species are 52% of the total of all known Galápagos species combined (Peck 1997). A "species-scape" shows the general relative species diversity of the major groups of terrestrial and marine Galápagos organisms (Fig. 1.8). Species numbers alone do not give a measure of the complete importance of a group in its entire ecosystem. In terrestrial habitats, the biomass of vascular plants certainly dominates their ecosystems. Even though vertebrates may be individually more conspicuous in their body size, I estimate that they are relatively of less total biomass and that insect biomass comes second after that of vascular plants. This has been calculated to be the case in temperate and tropical continental habitats (Wilson 1992).

The dominance of Galápagos terrestrial diversity over that in marine habitats certainly seems to be not as great as the global average. Why would this be? One obvious partial explanation is that the Galápagos terrestrial habitats are much younger that the marine ones. Another is that while the sea is a dispersal highway for many (but not all) marine organisms, it is a formidable barrier to

Fig. 1.8. A "species-scape" of the native multicellular terrestrial and marine organisms of the Galápagos Islands. Size of each "icon" organism is proportional to the known species diversity in the major taxon (Wheeler 1990). Unit area is equal to 10 species. Major taxa with less than 10 species are not indicated. Diversity data from various sources, summarized in Table 1.5 (and Peck 1997). Key to taxa. 1. Angiosperm plants. 2. Fungi. 3. Mosses and liverworts. 4. Ferns and allies. 5. Lichens. 6. Insecta (insects). 7. Aranea (spiders). 8. Mollusca (land snails). 9. Crustacea (terrestrial crustaceans). 10. Reptiles 11. Land birds. 12. Land mammals. 13. Centipedes and millipedes. 14. Acari (mites). 15. Nematoda. 16.Marine Crustacea. 17. Bryozoa. 18. Nematoda. 19. Mollusca (Bivalvia; clams). 20. Platyhelminthes. 21. Annelida. 22. Mollusca (Gastropoda; snails). 23. Marine "Algae." 24. Chelicerata. 25. Cnidaria (corals and jellyfish). 26. Echinoderms (starfish, sea urchins, brittle stars, etc.). 27. Fishes. 28. Sea birds. 29. Marine mammals. Figure by Jarmila Peck.

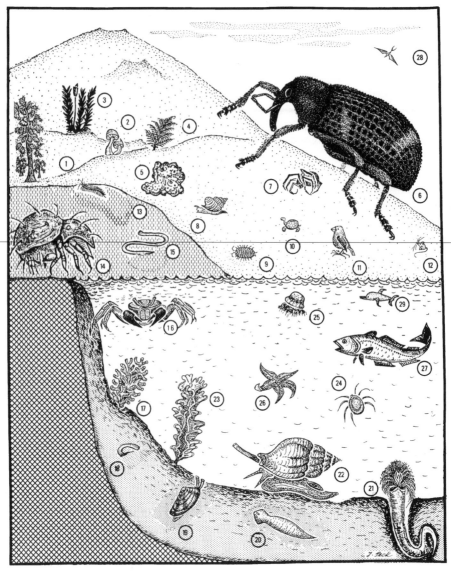

terrestrial organisms. Dispersal for terrestrial organisms to the islands must be through the air or on the surface of the sea. Another explanation is that, upon arrival, the habitat must be suitable for the colonists. This is more likely to be optimal in the sea than in the suboptimal and seasonally intensely arid conditions on land. Lastly, islands are generally known to have a smaller subset of species per unit area than continents because of their decreased habitat diversity. While the marine habitat diversity of the waters of the islands may be roughly equivalent to that available on the coasts of continental Ecuador, it is certainly not so for terrestrial habitats.

We can also see that some 53% of the terrestrial invertebrate species (1174 endemics in 2198 species) are endemics and that only some 32% of the marine invertebrate species (561 endemics in 1756 species) are endemics. This means that once the ancestral terrestrial colonists reached the islands they were much more likely to be isolated there and to evolve species differences. The higher percentage of endemics in the terrestrial invertebrate fauna is a direct consequence of the increased isolation of the terrestrial invertebrates once they reached the islands.

Biodiversity on other archipelagos

The only other tropical oceanic archipelago that I know of for which total biodiversity has been summarized is that of the Hawaiian Islands. These are the world's most isolated island group, and are larger and older than the Galápagos. Eldredge and Miller (1995, 1997) find that there are about 15 000 terrestrial, 300 freshwater, and 5500 marine species of plants and animals in Hawaii. Of the total, 21% are adventive and most of these are terrestrial insects. There is a high level of endemicity (41%) and most of this is in insects, plants, and land snails. These are patterns similar to those of the Galápagos. The islands of the Old World have been so long altered by human activities that it may no longer be possible to achieve estimates of their original terrestrial species diversities.

References

Adamson, A.M. 1939. Review of the fauna of the Marquesas Islands and discussion of its origin. Bernice P. Bishop Mus. Bull. **159**: 93 pp.

Ashmole, N. P., and Ashmole, M. J. 1997. The land fauna of Ascension Island: new data from caves and lava flows, and a reconstruction of the prehistoric ecosystem. J. Biogeogr. **24**: 549–589.

Baert, L. 1994. Notes on the status of terrestrial arthropods in Galápagos. Noticias de Galápagos no. **54**: 15–21.

Baert, L., Maelfait, J.-P., and Desender, K. 1989a. Results of the Belgian 1986 expedition: Araneae, and provisional checklist of the spiders of the Galápagos archipelago. Bull. Inst. R. Sci. Nat. Belg. Entomol. **58**: 29–54.

Baert, L., Maelfait, J.-P., and Desender, K. 1989*b*. Results of the Belgian 1988-expedition to the Galápagos Islands: Araneae. Bull. Inst. R. Sci. Nat. Belg. Entomol. **59**: 5–22.

Baert, L., Maelfait, J.-P., and Desender, K. 1995. Distribution of the arachnid species of the orders Scorpiones, Solifugae, Amblypygi, Schizomida, Opiliones and Pseudoscorpiones in Galápagos. Bull. Inst. R. Sci. Nat. Belg. Entomol. **65**: 5–19.

Baez, M. 1984. Los artropodos. *In* Fauna (marina y terrestre) del Archipelago Canario. *Edited by* J.J. Bacallado. Edirca, Las Palmas. pp. 101–254.

Baez, M. 1987. Charactères liés à l'insularite de la faune de l'archipel des Canaries. Bull. Soc. Zool. France **112**: 143–152.

Basiliwsky, P. 1985. South Atlantic island of St. Helena and the origin of its beetle fauna. *In* Taxonomy, phylogeny, and zoogeography of beetles and ants. *Edited by* G.E. Ball. W. Junk Publ., The Hague, Ser. Entomol. **33**: 257–275.

Basset, Y., Samuelson, G.A., Allison, A., and Miller, S.A. 1996. How many species of host-specific insects feed on a species of tropical tree? Biol. J. Linn. Soc. **59**: 201–216.

Becker, P. 1975. Island colonization by carnivorous and herbivorous Coleoptera. J. Anim. Ecol. **44**: 893–906.

Berry, R.J. (*Editor*). 1984. Evolution in the Galápagos. Academic Press, London. 260 pp. Reprinted from Biol. J. Linn. Soc. (London) 21: no. 1 and 2, 1984.

Borges, P.A.V. 1990. A checklist of the Coleoptera from the Azores with some systematic and biogeographic comments. Biol. Mus. Mun. Funchal **42**: 87–136.

Bowman, R.I. (*Editor*). 1966. The Galápagos; proceedings of the Galápagos International Scientific Project. Univ. Calif. Press, Berkeley, CA, 318 pp.

Bowman, R.I., Benson, M.,and Leviton, A.E. (*Editors*). 1983. Patterns of evolution in Galápagos organisms. Pacific Division, AAAS, California Acad. Sci., San Francisco, CA, 568 pp.

Buxton, P.A. 1935. Insects of Samoa. British Museum (Nat. Hist.) Part IX. Fasc. 2, Summary 33–104.

Carlquist, S. 1965. Island life. Natural History Press, Garden City, New York. viii–451 pp. [References to Galápagos insects on pp. 180, 181, 218, 219, 224].

Carlquist, S. 1974. Island biology. Columbia University Press, New York, NY, 660 pp.

Cheesman, L.E. 1927. A contribution towards the insect fauna of French Oceania. Trans. Entomol. Soc. London **75**: 147–161.

Cheesman, L.E. 1934. The insect fauna of French Polynesia. Soc. Biogeog. **4**: 191–200.

Christie, D.M., Duncan, R.A., McBirney, A.R., Richards, M.A., White, W.M., Harpp, K.S., and Fox, C.G. 1992. Drowned islands downstream from the Galápagos hotspot imply extended speciation times. Nature **355**: 246–248.

Cogan, B.H., Hutson, A.M., and Schaffer, J.C. 1971. Preliminary observations on the affinities and composition of the insect fauna of Aldabra. Phil. Trans. R. Soc. Lond. B **260**: 315–325.

Colinvaux, P.A. 1972. Climate and the Galápagos Islands. Nature **240**: 17–20.

Colinvaux, P.A. 1984. The Galápagos climate: present and past. *In* Galápagos. Key Environments. *Edited by* R. Perry. Pergamon Press, New York. pp. 55–69.

Colinvaux, P.A. and Schofield, E.K. 1976. Historical ecology in the Galápagos Islands. I. A Holocene pollen record from El Junco Lake, Isla San Cristóbal. J. Ecol. **64**: 989–1012.

Cox, A. 1983. Ages of the Galápagos Islands. *In* Patterns of evolution in Galápagos organisms. *Edited by* R.I. Bowman, M. Berson, and A.E. Leviton. AAAS, Pacific Division, Calif. Acad. Sci., San Francisco, CA. pp. 11–23.

Desender, K., and Baert, L. 1996. The Coleoptera of Easter Island. Bull. Inst. R. Sci. Nat. Belg., Entomol. **66**: 27–50.

Desender, K., Baert, L., Maelfait, J.-P., and Verdyck, P. 1999. Conservation on Volcán Alcedo (Galápagos): terrestrial invertebrates and the impact of introduced feral goats. Biol. Conserv. **87**: 303–310.

De Vries, T.J., Ortlieb, L., Diaz, A., Wells, L., and Hillaire-Marcel, Cl. 1997. Determining the early history of El Niño. Science **276**: 965–967.

Duncan, R.A., and Hargraves, R.B. 1984. Plate tectonic evolution of the Caribbean region in the mantle reference frame. Geol. Soc. Am. Mem. **162**: 81–93.

Eldredge, L.G., and Miller, S.E. 1995. How many species are there in Hawaii? Bishop Mus. Occas. Pap. **41**: 1–18.

Eldredge, L.G., and Miller, S.E. 1997. Numbers of Hawaiian species: supplement 2, including a review of freshwater invertebrates. Bishop Museum Occ. Pap. **48**: 3–22.

Erber, D., and Hinterseher, W. 1988. Contribution to the knowledge of the Madeira beetles. Bol. Mus. Mun. Funchal. **40**: 139–214.

Eliasson, U. 1984. Indigenous climax forests. *In* Key Environments. Galápagos. *Edited by* R. Perry. Pergamon Press, New York. pp. 101–114.

Finston, T., and Peck, S.B. 1997. Genetic differentiation and speciation in *Stomion* (Coleoptera: Tenebrionidae): flightless beetles of the Galápagos Islands, Ecuador. Biol. J. Linn. Soc. (London) **61**: 183–200.

Finston, T., Peck, S.B., and Perry, R.B. 1997. Population density and dispersal ability in Darwin's Darklings: flightless beetles of the Galápagos Islands, Ecuador. Pan-Pac. Entomol. **73**: 110–121.

Gertsch, W., and Peck, S.B. 1992. The pholcid spiders of the Galápagos Islands, Ecuador (Araneae: Pholcidae). Can. J. Zool. **70**: 1185–1199.

Geist, D. 1996. On the emergence and submergence of the Galápagos Islands. Noticias de Galàpagos **56**: 5–9.

Geisthardt, M. 1988. Tabellarische Ubersicht zur Verbreitung der Coleoptera auf den Kapverdischen Inseln. Cour. Forsch.-Inst. Senckenberg **105**: 193–210.

Gillerfors, G. 1986. Contribution to the Coleopterous fauna of the Azores. Bol. Mus. Mun. Funchal **38**: 16–27.

Graham, J.B. 1975. The biological investigation of Mal Pelo Island, Columbia. Smithson. Contrib. Zool. **176**: 1–98.

Grant, P.R. 1986. Ecology and evolution of Darwin's finches. Princeton Univ. Press, Princeton, NJ. 458 pp.

Grant, B.R., and Grant, P.R. 1989. Evolutionary dynamics of a natural population: the large cactus finch of the Galápagos. Univ. Chicago Press, Chicago, IL, 350 pp.

Gressitt, J.L. 1961. Problems in the zoogeography of Pacific and Antarctic insects. Pac. Insects Monogr. **2**: 1–94.

Hamann, O. 1993. The vegetation of Isla Santiago – past and present. Noticias de Galápagos **52**: 6–11.

Hauff, R., Hoernle, K., Schmincke, H.-U., and Werner, R. 1997. Geol. Rundsch. **86**: 141–155.

Hillburn, D.J., and Gordon, R.D. 1989. Coleoptera of Bermuda. Fla. Entomol. **72**: 673–692.

Hogue, C.L., and Miller, S.E. 1981. Entomofauna of Cocos Island, Costa Rica. Atoll Res. Bull. **250**: 29 pp + additions 4 pp.

Howarth, F.G. 1990. Hawaiian terrestrial arthropods: an overview. Bishop Mus. Occas. Pap. **30**: 4–26 (Honolulu).

Itow, S. 1995. Phytogeography and ecology of *Scalesia* (Compositae) endemic to the Galápagos Islands. Pac. Sci. **49**: 17–30.

Jackson, M.H. 1985. Galápagos, a natural history guide. University of Calgary Press, Calgary, AB, 283 pp.

Jackson, J.B.C., Budd, A.F., and Coates, A.G. 1996. Evolution and environment in tropical America. Univ. Chicago Press, Chicago, IL.

James, M.J. (*Editor*). 1991. Galápagos marine invertebrates; taxonomy, biogeography, and evolution in Darwin's Islands. Topics in Geobiology vol. 8. Plenum Press, New York.

Kuschel, G. 1963. Composition and relationship of the terrestrial faunas of Easter, Juan Fernandez, Desaventuradas, and Galápagos Islands. Occas. Pap. Calif. Acad. Sci. **44**: 79–95.

Lawton, J.H., and Schroder, D. 1977. Effects of plant type, size of graphical range and taxonomic isolation on number of insect species associated with British plants. Nature **265**: 137–140.

Lindberg, H. 1963. A contribution to the study of beetles in the Madeira Islands. Soc. Sci. Fenn. Comm. Biol. **25**(2): 1–156.

Linsley, E.G. 1977. Insects of the Galápagos (Supplement). Occas. Pap. Calif. Acad. Sci. **125**: 1–50.

Linsley, E.G., and Usinger, R.L. 1966. Insects of the Galápagos Islands. Proc. Cal. Acad. Sci. **33**: 113–196.

Lundblad, O. 1958. Die Arthropodenfauna von Madeira nach den Ergebnissen der Reise von Prof. Dr. O. Lundblad Juli-August, 1935. XXXV. Die Kaferfauna der Inseln Madeira. Ark. Zool. (2) 11 (3 =): 461–524.

Mann, P. 1995. Preface. Geologic and tectonic development of the Caribbean Plate Boundary in southern Central America. Geol. Soc. Am. Spec. Pap. 295. pp. i–xxxi.

Méquignon, A. 1946. Le peuplement entomologique des Acores. Soc. Biogeog. Mem. **8**: 109–134.

Oromí, P. 1983. Sobre el origin de la fauna entomologica de las islas Salvages. Vieraea **12**: 271–293.

Paulian, R. 1961. La zoogéographie de Madagascar et des iles voisines. Faune Madagascar XIII.

Peake, J.F. 1971. The evolution of the terrestrial faunas in the western Indian Ocean. Phil. Trans. Roy. Soc. Lond. B **260**: 581–610.

Peck, S.B. 1991. The Galápagos Archipelago, Ecuador: with an emphasis on terrestrial invertebrates, especially insects; and an outline for research. *In* The unity of evolutionary biology. Proceedings of the Fourth International Congress of Systematic and Evolutionary Biology. *Edited by* E.C. Dudley. Dioscorides Press, Portland, OR. pp. 319–336.

Peck, S.B. 1993. Galápagos species diversity: is it on the land or in the sea? Noticias de Galápagos **52**: 18–21.

Peck, S.B. 1994. Diversity and zoogeography of the non-oceanic Crustacea of the Galápagos Islands, Ecuador (excluding terrestrial isopoda). Can. J. Zool. **72**: 54–69.

Peck, S.B. 1997. The species-scape of Galápagos organisms. Noticias de Galápagos **58**: 18–21.

Peck, S.B., and Shear, W.A. 2000. New records of Myriapoda (centipedes and millipedes) from the Galápagos Islands. Noticias de Galápagos No. **61**: 14–16.

Peck, S.B. and Shear, W.A. 2001. New records of centipedes and millipedes (Arthropoda: Tracheata: Chilopoda and Diplopoda) from the Galapagos Islands. Noticias de Galapagos. In press.

Peña, L. 1987. Consideraciones sobre la fauna de arthropodos terrestres de las Islas Oceanicas Chilenas. In Islas Oceanicas Chilenas: Conocimiento cientifico y necesidades de investigaciones. *Edited by* J.C. Castilla. Ediciones Universidad Catolica de Chile. pp. 217–223.

Perry, R. 1984. Galápagos. Key Environments. Pergamon Press, New York. 336 pp.

Reeder, W.C., and S.E. Reichardt. 1975. Vegetation along an altitudinal gradient, Santa Cruz, Galápagos Islands. Biotropica **7**: 162–175.

Rosen, D.E. 1975. A vicariance model of Caribbean biogeography. Syst. Zool. **24**: 431–469.

Sandweiss, D.M., Richardson III, J.B., Reitz, E.J., Rollins, H.B., and Maasch, K.A. 1996. Geoarcheological evidence from Peru for a 5000 years B.P. onset of El Niño. Science **273**: 1531–1533.

Schatz, H. 1991. Catalogue of known species of Acari from the Galápagos Islands (Ecuador, Pacific Ocean). Int. J. Acarol. **17**: 213–225.

Schatz, H. 1998. Oribatid mites from the Galápagos Islands — faunistics, ecology and speciation. Exp. Appl. Acarology **22**: 373–409.

Scott, H. 1933. General conclusions regarding the insect fauna of the Seychelles and adjacent islands. Trans. Linn. Soc. London, 2nd Ser., Zool. **19**: 307–391.

Serrano, A.R.M. 1982. Contribuçao para o conhecimento do povoamento, distribiçao e origem dos Coleopteros do Arquipelago dos Açores (Insects, Coleoptera). Bolm. Mus. Mun. Funchal **34**(147): 67–104.

Shear, W.A., and Peck, S.B. 1987. Millipeds (Diplopoda) of the Galápagos Islands, Ecuador. Can. J. Zool. **65**: 2640–2645.

Shear, W.A., and Peck, S.B. 1992. Centipeds (Chilopoda) and Symphyla of the Galápagos Islands, Ecuador. Can. J. Zool. **70**: 2260–2274.

Slevin, J.R. 1959. The Galápagos Islands: a history of their exploration. Occas. Pap. Calif. Acad. Sci. **25**: 1–150.

Smith, K.G.V. 1987. Darwin's insects. Bull. Br. Mus. Nat. Hist. (Hist. Ser.) **14**: 1–143.

Snell, H.M., Stone, P.A., and Snell, H.L. 1995. Geographical characteristics of the Galápagos Islands. Noticias de Galápagos no. **55**: 18–24.

Snell, H.M., Stone, P.A., and Snell, H.L. 1996*a*. A summary of geographical characteristics of the Galápagos Islands. J. Biogeogr. **23**: 619–624.

Snell, H. M., Snell, H. L., Davis-Merlen, G., Simkin, T., and Silberglied, R. E. 1996*b*. Bibliografía de Galápagos 1535–1995. Fundación Charles Darwin para las Islas Galápagos. Quito, Ecuador.

Southwood, T.R.E. 1960. The abundance of the Hawaiian trees and the number of their associated insect species. Proc. Hawaii Entomol. Soc. **17**: 299–303.

Southwood, T.R.E. 1961. The number of species of insect associated with various trees. J. Anim. Ecol. **30**: 1–8.

Strong, D.R., Jr., and Levin, D.A. 1979. Species richness of plant parasites and growth form of their hosts. Am. Nat. **114**: 1–22.

Strong, D.R., Jr., Lawton, J.H., and Southwood, T.R.E. 1984. Insects on plants. Harvard Univ. Press, Harvard, MA.

Sulloway, F.J. 1984. Darwin and the Galápagos. Biol. J. Linn. Soc. **21**: 29–59.

Swezey, O.H. (*Editor*). 1942. Insects of Guam I. Bernice P. Bishop Mus. Bull. **172**: 1–218.

Swezey, O.H. (*Editor*). 1946. Insects of Guam. II. Bernice P. Bishop Mus. Bull. **189**: 1–237.

Thornton, I.W.B. 1971. Darwin's Islands, a natural history of the Galápagos. Natural History Press, Garden City, NY. 322 pp.

Usinger, R.L. 1972. Robert Leslie Usinger: autobiography of an entomologist. Mem. Pac. Coast Entomol. Soc. vol 4. 330 pp.

Uyttenboogaart, D.L. 1946. Le peuplement des iles Atlantides. Conclusions a tirer de la composition de la fauna des coléopteres. Soc. Biogeog. Mem. **8**: 135–152.

Wheeler, Q. 1990. Insect diversity and cladistic constraints. Ann. Entomol. Soc. Am. **83**: 1031–1047.

White, W.M., McBirney, A.R., and Duncan, R.A. 1993. Petrology and geochemistry of the Galápagos Islands: portrait of a pathological mantle plume. J. Geophys. Res. **98**: 19533–19563.

Wiggins, I.L., and Porter, D.M. 1971. Flora of the Galápagos Islands. Stanford Univ. Press, Stanford, CA. 998 pp.

Wilson, E.O. 1992. The diversity of life. W.W. Norton & Co., NY.

Wollaston, T.V. 1857. Catalogue of the coleopterous insects of Madeira in the collection of the British Museum. Taylor and Francis, London. 234 pp.

Wollaston, T.V. 1864. Catalogue of the coleopterous insects of the Canaries, in the collections of the British Museum. Taylor and Francis, London. 548 pp.

Wollaston, T.V. 1865. Coleoptera Atlantidum, being an enumeration of the Coleopterous insects of the Maderias, Salvages and Canaries. Taylor and Francis, London. 526 pp + 140 pp appendix.

Wyles, J.S., and Sarich, V.M. 1983. Are the Galápagos iguanas older than the Galápagos? Molecular evolution and colonization models for the archipelago. *In* Patterns in evolution in Galápagos Organisms. *Edited by* R.I. Bowman, M. Benson, and A.E. Leviton. Pacific Division, AAAS, San Francisco, CA, pp. 177–186.

Zimmerman, E.C. 1948. Insects of Hawaii. Introduction. Univ. Hawaii Press, Honolulu, Hawaii. Vol. 1. 206 pp.

Zimmerman, E.C. 1970. Adaptive radiation in Hawaii with special reference to insects. Biotropica **2**: 32–38.

Chapter 2
Information sources about Galápagos
Insects

Previously published information

There is a very extensive literature on Galápagos insects. By 1977 there had been 23 multidisciplinary collecting programs that had gathered insect data, and information on 883 species had appeared in 308 scientific publications. The specimens resulting from these historical collections are probably mostly in the museums or institutions originally sponsoring the expeditions. The only exception known to me are the specimens from the 1898–1899 Hopkins expedition. The entire insect collection of Stanford University was disassembled in the 1960's and the fate of the Galápagos specimens is not fully known. Most specimens seem to have gone to the Los Angeles County Museum, but at least some are in the California Academy of Sciences, including some type specimens. The data for my analyses in Chapters 3–5 and species summaries in Chapters 6–23 have been compiled from the 308 scientific references, as given by Linsley and Usinger (1966) and Linsley (1977), and in papers published since these summaries. Additional data have come from unstudied or unpublished collections in various museums (see below), and from the collections of my field teams. The lists of Linsley and Usinger (1966) and Linsley (1977) give complete citations for adventive or indigenous species names from the older literature. These citations have not been repeated here in the references. With author name, year and page of publication it is possible to locate these older references in the Zoological Record or in general bibliographies which do not deal directly with Galápagos records.

Museum collections

Many new records were gained from previously unstudied or unpublished specimens in the collections of several museums. These were the Charles Darwin Research Station (CDRS, Isla Santa Cruz, Galápagos, Ecuador), the Bernice P. Bishop Museum (BPBM, Honolulu, HI), the California Academy of Sciences (CASC, San Francisco, CA), the Museum of Comparative Zoology (MCZ, Harvard University, Cambridge, MA), Department of Entomology, the U.S. National Museum of Natural History (USNM, Smithsonian Institution, Washington, D.C), and Koninklijk Belgisch Instituut voor Natuurwetenschappen (KBIN, Brussels, Belgium).

Field methods

The bulk of the new information in this book is from the field work of myself and members of my field teams from June–August 1985; January–April 1989; April–July 1991; March–May 1992; and March–May 1996. Collections were mostly made during general insect survey field work. The sampling methods which were used were many. They included: (1) hand searching under rocks, logs, and in leaf litter; (2) searching, beating, and sweeping of vegetation during both nighttime and daytime; (3) the use of baited and unbaited pitfall traps, bottle traps, and pit traps set in deep covered holes in soil and ash, Malaise traps, yellow pan traps, and flight-intercept traps (FIT); (4) collecting at night at ultraviolet and white lights; (5) net collecting of day-active specimens, and (6) collecting in special habitats and by special methods such as along "oatmeal trails" at night, in caves, in bird nests, Berlese extraction of leaf litter and soil, etc. These methods were used in villages, in disturbed agricultural habitats, and extensively in all possible natural (undisturbed) habitats and all six major "life zones" (Fig. 1.5) of as many islands as could be visited. Collecting efforts of my teams are summarized in Tables 2.1, 2.2, and 2.3. All major islands and most minor islands have been sampled at least once, including the remote islands of Darwin and Wolf in the extreme northwestern corner of the Archipelago.

We attempted to give sampling attention to islands in proportion to their size and ecological complexity. However, the difficulties of logistics, high costs involved in reaching islands distant from the Charles Darwin Research Station, and vageries of rainfall on individual islands did not always result in the desired equivalency of sampling. For example, San Cristóbal, Española, and Floreana are probably still less well sampled when compared to the other large islands. Santa Cruz is certainly the best sampled island because it is the location of the logistical base of the Charles Darwin Research Station.

I expect that additional and specialized collecting techniques will continue to find some new species and new island records for known species. But still, I estimate that 90–95% of the insect species have now been collected (but all are not yet identified as either named or new species). Although this data base is thus not entirely complete, I believe that the generalities presented here for the evolutionary and ecological patterns will not significantly change. The goal of a more complete inventory is offered as a challenge to future students of Galápagos insects. No other archipelago in the world can now offer such a research opportunity into the natural and relatively undisturbed composition and structure of a tropical oceanic insect assemblage. Most of the world's other island insect faunas were markedly changed by humans before adequate scientific sampling was possible. The growing numbers and spread of adventive insect species in the Galápagos suggest that even the results of the present study may not be fully repeatable in the future.

Table 2.1. Summary of my teams' effort for sampling insects on the Galápagos Islands as total numbers of sampling stations per principal collector. Several or many sampling methods or traps may have been at each station.

	1985	1989	1991	1992	1996
S.B. Peck	218	212	186	230	207
B. Landry (Lepidoptera)		24		57	
B. Sinclair (Diptera)		193			
J. Heraty (Hymenoptera)			127		
R. Palma (Phthiraptera)				590	
Subtotals	218	429	313	877	207

Grand total = 2044 total insect sampling stations

Organization of the inventory

The insect orders are listed here in a generally accepted phylogenetic sequence. A summary total is given for each order of the number of families, genera, number of species (categorized as endemic, indigenous, or adventive), and an estimate of the numbers of natural ancestral colonizations. Within an order the suborders or families are listed in a generally used sequence, often following that used in textbooks such as Borror et al. (1989), unless specialists or recent publications have suggested another sequence. Genera and species are listed in alphabetical order.

Keys

Keys that may be used to identify Galápagos insects to order or family are in Borror et al. (1989) or Arnett (1993) and are not repeated here for the large or difficult orders. Additional information on Neotropical insects is in Hogue (1993). I have constructed keys for the identification of families, genera, or species, where possible, for the use of future field workers and students and to allow them to try to make quick determinations of Galápagos insects in groups in which they are not a specialist. Keys made by others have been credited to the proper author. It was not possible to provide keys for some small-bodied insect groups where specialist knowledge and microscopic study of slide prepared specimens is required, such as for the aphids, leafhoppers, scale insects, thrips, and bark lice.

Specimens

A small reference collection of identified specimens of many of the conspicuous or common species is deposited the the museum of the Darwin Station, and

Table 2.2. Summary of total insect sampling effort by island and life zone of my teams in the Galápagos Islands, 1985–1996, as combined total number of samples, days of sampling, and days of trapping per island and per life zone.

Island	Littoral	Arid	Transition	Humid Forest	Evergreen Shrub	Pampa	Inversion	Urban	Agriculture	Total
Baltra	14	17								31
Bartolomé	5	3								8
Campeón	4	5								9
Caamaño	16	3								19
Cowley		3								3
Darwin	1	10								11
Española	50	178								228
Fernandina	15	316	54			5				395
Floreana	88	111	78	265					15	557
Gardner at Española	2	5								7
Gardner at Floreana	2	5								7
Genovesa	7	146								153
Isabela	319	334	110	11		206	46	20	102	1148
Marchena	1	147								148
Pinta	21	189	21	21						252
Pinzón		57		1		2				60
Plazas Sur		4								4
Rábida	12	109								121
San Cristóbal	147	106	49	6	27	70		6	54	465
Santa Cruz	432	1579	844	769	453	651		3	314	5045
Santa Fé		15	88							103
Santiago	8	144	35	140	39	46				412

Table 2.2 (*concluded*).

Island	Littoral	Arid	Transition	Humid Forest	Evergreen Shrub	Pampa	Inversion	Urban	Agriculture	Total
Seymour Norte	5	17								22
Tortuga		6								6
Wolf	1	88								89

Vegetation Life Zones

Table 2.3. Summary of total insect sampling effort by island and sampling technique of my teams in the Galápagos Islands, 1985–1996, as combined number of samples and (or) days of trapping. Totals on this table do not match those on Table 2.2 because all of the sampling methods that were used are not shown on this table.

Island	Sampling method											Vertebrate Hosts	Totals
	General	Litter	Sweeping	Beating	Sea Cliff Spraying	Light Traps	Flight Traps	Malaise Traps	Yellow Pans	Pit Traps	Caves		
Baltra	8		1		2			20					31
Bartolomé	6	1			1								8
Campeón	6			2	1								9
Caamaño	5					7	7						19
Cowley	3												3
Darwin	6		2	1	1	1							11
Española	6	8	6	5	2	9	14	14		52		109	225
Fernandina	3	7	19	9	5	5	11	11	6	300		17	393
Floreana	16	17	14	6		16	144	144	35	161	3	14	570
Gardner at Española	7												7
Gardner at Floreana	7												7
Genovesa	4	2		1	1	5	17	17	60	1	42		150
Isabela	43	45	57	15	6	49	92	107	278	423	16	5	1136
Marchena	3	2			3		25	25		73		17	148
Pinta	6	5	1	6	1	9	25	25	6	17	2	140	243
Pinzón	2	3	7						48				60
Plazas Sur	2	2											4
Rábida	4	5	5		1	3	18	18	54		12		120
San Cristóbal	16	28	5	6	1	22	79	79	157	59			452
Santa Cruz	47	64	41	21	1	66	1118	985	189	1704	695	101	5032

Table 2.3 (*concluded*).

Island	Sampling method												
	General	Litter	Sweeping	Beating	Sea Cliff Spraying	Light Traps	Flight Traps	Malaise Traps	Yellow Pans	Pit Traps	Caves	Vertebrate Hosts	Totals
Santa Fé	3		6	2	3	2	2	13	22	28	22		103
Santiago		8	11	12		18	41	48	9	168		90	405
Seymour Norte	6	2		1	4				3		6		22
Tortuga	5	1											6
Wolf	3		2		2	1	1		80				89

Vertebrate Hosts: Total individuals searched: 575; individuals with ectoparasites, 240; individuals without ectoparasites, 235.

Inter-island aerial and sea surface sampling: Aerial: 27 samples, 161 h of sampling; 1 820 000 m^3 of air; Sea surface: 18 samples; 42 h of sampling; 130 000 m^2 of sea surface.

the insect collection of the Pontifical Catholic University of Ecuador (PUCE) in Quito. My synoptic reference collection and voucher specimens are deposited in the Entomology Department, U.S. National Museum of Natural History (USNM). Deposition of other specialty collections are indicated in the appropriate chapters. Residues of samples are in collections of the Field Museum, Chicago, IL (FMNH).

Synonomy and citations

An abbreviated synonomy for each species is given, and is as complete as is judged to be useful for names applied only to taxa in the Galápagos and elsewhere in Latin America. Synonomies used only in an Old World context are not given, but can usually be found in general catalogues. Author name of each species is followed by year and page of publication when I have been able to find it. Full publication data are not given in the "literature cited" for all these references, but only for those that help understand the Galápagos fauna. For example, references are not given for older authors such as Linnaeus, Fabricius, Gyllenhal, etc. These are usually in Linsley and Usinger (1966) and Linsley (1977).

Many Galápagos insect species have been mentioned in many publications. I cite here only those that contribute new information on a species, not catalogues or lists that give no new information. These references are given after the species names in order of year of publicaton, not in alphabetical order by author.

Distributions

A summary hypothesis of residency status and distribution is given for each species. Galápagos species are placed into three categories as follows. (1) *Adventive* species were probably unintentionally brought to the islands in historical time by *human agency or transport*. These have frequently been called "introduced" species, but this can imply both intentional and accidental human transport. There is no evidence that any insects have been intentionally carried to the Galápagos. (2) *Indigenous* species are those thought to *naturally* occur both in the Galápagos and elsewhere (usually Latin America). They have arrived in the Galápagos in relatively recent evolutionary time and have not taxonomically differentiated from their populations living beyond the islands. In most earlier Galápagos literature these species have been called "native," but this term is used here with another long-established meaning (see below). (3) *Endemic* species are *naturally limited to one or more of the islands* in the Archipelago. Their ancestors arrived long enough ago that they have taxonomically differentiated (speciated) from their source populations. Species and genera which are taxonomically very remote from posssible ancestors may represent exceptionally early colonizations and can be called *paleo-endemics*. There are a few examples of these in Galápagos insects.

The "adventive" category is the most subjective because of the lack of documentation of insect introductions, either intentional or accidental. Generally, species known to be associated with human activities, human-altered habitats, stored products, construction and agricultural materials, and agricultural or horticultural plants or animals are judged to be adventive (Peck et al. 1998). If a species is not judged to be adventive, and occurs outside of the archipelago it is then thought to have naturally dispersed to the Galápagos and is classed as an indigenous species.

To contrast them with those species which were probably transported by human agency, all the naturally occurring species (the *indigenous and endemic species combined*) can be grouped together and called *native* species. I have used the above terms as they have been generally applied to the Hawaiian insect faunas (Nishida 1994). There is an extensive literature of alternate terms (see Frank and McCoy 1990 and Zimmerman 1948), but they are not useful here in understanding the status or origins of the Galápagos insect fauna.

Distributional information and zoogeographic analysis very much depends on the accuracy of the species determinations. Future workers may eventually discover that species described as endemics may later be found to occur on the South or Central American mainland. The distributional data given here for the indigenous species of insects is drawn from many various sources, and may be incomplete. Numbers of ancestral colonizations are estimated by judging if a species' sister (or sisters) is (are) in the Archipelago (both (all) are products of a single ancestral colonization) or if the sister occurs elsewhere than the Galápagos. Few Galápagos species have been subjected to a rigorous cladistic analysis to verify such estimates of numbers of ancestral colonizations.

Island names

Many of the Galápagos islands have traditional English names, and many of these are the oldest available name for an island. But the rules of taxonomic priority do not apply here. To attempt to avoid confusion, I use here only the preferred Ecuadorian names for the islands. I think it more appropriate to use the names as recognized by the country to which they belong and by which they are administered, and to suppress the use of the English names. These preferred Ecuadorian names are listed in Table 1.1, which also includes the English names and other synonymies of island names. If older literature records use the English names, they are converted here to the preferred Ecuadorian names.

Bionomics

Summary bionomic information is given on the general feeding habits or habitats of the species, the vegetation life zones they are known to inhabit, seasonality of collections of adults, special methods of collection, and host plant records (if known). Data are from published literature, specimen labels, or field observations. Feeding habits are not always based on direct observation. There is remarkably little direct evidence about the feeding habits of Galápagos insects. I use the following trophic categories:

Predators – These are species in which larvae and adults principally feed on living animal tissues (carnivores).

Scavengers – These are species in which larvae and adults principally feed on dry or moist dead organic matter (saprophages). Some specialize in dead plant matter, and others specialize on dead animal tissues or wastes. It does not seem helpful at present to distinguish between these subdivisions because most Galápagos scavengers are omnivorous, feeding on both types of substances.

Fungivores – These are species in which larvae and adults principally feed on dry or moist fungal tissues.

Herbivores – These are species in which larvae and adults principally feed on living plant tissue, such as plant fluids and juices, seeds and soft or hard parts of vascular plants. Included here are species that feed on living sapwood, and that bore into living sapwood or stems. Subcategories include wood and stem borers, leaf feeders, leaf miners, fruit and seed predators, sap and juice feeders, and pollen and nectar feeders. In the small orders, there are no known leaf miners in the Galápagos.

Parasites – These are species in which larvae (rarely adults) principally feed on or in a single host animal, usually an immature insect or a terrestrial vertebrate. They may be endoparasites (Strepsiptera) or ectoparasites (fleas). No parasitoids are known in the small orders of Galápagos insects.

Summaries

The above data on taxonomy, distribution, and bionomics are used to form the summaries of evolutionary processes and ecological patterns which follow in Chapters 3, 4, and 5.

References

Arnett, R.M. Jr. 1993. American insects: a handbook of the insects of America north of Mexico. Sandhill Crane Press, Gainesville, FL.

Borror, D.J., Triplehorn, C.A., and Johnson, N.F. 1989. An introduction to the study of insects. 6th Edition. Saunders College Publ., Holt, Rinehart and Winston, Philadelphia, PA.

Frank, J.H., and McCoy, E.D. 1990. Endemics and epidemics of shiboleths and other things causing chaos. Fla. Entomol. **73**: 1–9.

Hogue, C.L. 1993. Latin American insects and entomology. University California Press, Berkeley, CA.

Linsley E.G. 1977. Insects of the Galápagos (Supplement). Occas. Pap. Cal. Acad. Sci. **125**: 1–50.

Linsley, E.G., and Usinger, R.L. 1966. Insects of the Galápagos Islands. Proc. Cal. Acad. Sci. **33**: 113–196.

Nishida, G. 1994. Hawaiian terrestrial arthropod checklist. Bishop Museum Tech. Rep. 4.

Peck, S.B., Heraty, J., Landry, B., and Sinclair, B.J. 1998. The introduced insect fauna of an oceanic Archipelago: The Galápagos Islands, Ecuador. Am. Entomol. **44**: 218–237.

Zimmerman, E.C. 1948. Insects of Hawaii. Introduction. University of Hawaii Press, Honolulu, HI. Vol. 1. 206 pp.

Chapter 3
Processes: origin and arrival of the insect fauna

The processes of evolution can seldomly be observed. Usually, they are deduced from an analysis of distributional and ecological patterns in conjunction with evolutionary theory. Thus, an analysis and discussion of these processes should logically come after a documentation of the species which are present, and an analysis of their ecological structure and distributional patterns. I have chosen not to use this sequence. Instead, because the colonization and evolutionary processes have constructed the fauna that is actually present today, I first consider these processes in this and the next chapter. These two chapters are based on an earlier summary (Peck 1996b) but are modified and updated. They consider the entire terrestrial insect and arthropod fauna. Then I consider the resultant ecological patterns (Chapter 5, which focuses on the smaller insect orders). The smaller orders represent 27% of the presently known species. How the ecological patterns may be different in the orders not covered here remains to be analysed. Lastly, I present the actual data in the species inventories for the smaller orders (Chapters 6–23).

Taxonomic composition of the insect fauna

A summary of known Galápagos species diversity for the terrestrial and freshwater arthropod faunas is given in Tables 1.4 and 1.5. Inspection of these tables shows several obvious facts and generalizations.

An unbalanced fauna. When compared to the fauna of the Neotropical continental source area, it is evident that the Galápagos fauna is impoverished and unbalanced (or disharmonic). This means that the taxonomic composition is significantly different in its makeup and proportions. This is true for the terrestrial invertebrates as well as the vertebrates and plants. The Oribatida mites are certainly unbalanced in their taxonomic composition (Schatz 1998).

The diversity, ecology, and vagility of the insect orders in the Neotropical source areas has an influence upon which of them could and could not arrive and survive on the Galápagos. Only a part of the diversity of the mainland orders and a very limited subset of the families have been able to disperse to and colonize the archipelago. Within the insects, most orders are present, but the following are unknown or absent: Protura, Ephemeroptera (mayflies), Plecoptera (stoneflies), Grylloblattodea (rock crawlers), Phasmatodea (walking sticks),

Rhaphidioptera (snake flies), Megaloptera (alderflies and dobsonflies), Mecoptera (scorpion flies), and Trichoptera (caddisflies).

The probable reasons for the absence of these are diverse. The special sampling needed to discover Protura has not been attempted. Rhaphidioptera, Grylloblattodea, and Mecoptera are mainly northern and cool-temperate terrestrial groups, and are either uncommon or absent in the Neotropical fauna. Phasmatodea are diverse in the Neotropics, but their general rarity on remote islands everywhere seemingly suggests that they are not good over-water dispersers. And the remaining four orders (Ephemeroptera, Plecoptera, Trichoptera, Megaloptera) are absent partly because of their low vagility and their ecological requirements: they need standing or running water for their larval development and these habitats are mostly absent or only seasonally present in the Galápagos (Gerecke et al. 1995). These four orders are usually absent or only poorly represented in the lowland tropical source faunas of Central and South America, and from oceanic islands in general.

Small orders such as Archeognatha (bristletails), Mantodea (mantids), Embioptera (web spinners), Zoraptera (zorapterans), Strepsiptera (twisted-wing parasites), and Siphonaptera (fleas) are present with one or only a few species.

The dominant orders in South America are also the dominant orders in the Galápagos; these are the Hemiptera (true bugs), Homoptera (leafhoppers, plant hoppers, scale insects, and aphids), Coleoptera (beetles), Diptera (flies), Lepidoptera (moths), and Hymenoptera (bees, ants and wasps). But in these diverse orders the unbalanced nature of the fauna is very evident. For example, in the Homoptera, the absent representatives are the cicada, treehopper, and spittlebug groups of families. The fulgoroid (plant hopper) and coccoid (scale insects) groups are represented by half or fewer of the Neotropical families. In Coleoptera, the large families in the Galápagos are Carabidae (predaceousground beetles), Staphylinidae (rove-beetles) and Curculionidae (weevils), as is generally true in the Neotropics. But many large and important subfamilies or tribes of these beetles are absent. Leaf-feeding scarab and chrysomelid beetles, and dung-feeding scarabs are absent or remarkably poorly represented. Tenebrionids (darkling ground beetles) make up a larger component of the fauna than in the Neotropics in general, because of the arid nature of the Galápagos lowlands, which favors such drought tolerant beetles. Likewise, most family groups of aquatic Hemiptera and Diptera are absent. The few aquatic insects which are present are some highly vagile dragonflies and a damselfly (Peck 1992), a few aquatic true bugs, and some water beetles, and these are mostly species which are tolerant of brackish coastal waters (Gerecke et al. 1995). Within the Lepidoptera, the butterfly fauna is very impoverished, and contains only 10 species in 4 families. These same families in the Neotropics contain over 7434 species. Mainland Ecuador is known to have at least 2000 species of butterflies (and perhaps as many as 3000 species) (Moreno Espinosa et al. 1997). With a mainland area of 275 679 km^2, this is a minimum density of 0.00725 species per square km on the mainland, compared to a much lower species density of 0.00126 species per square km on the islands. Bees and wasps and ants are also conspicuous in either their family absence or very low species diversity. A mainland Neotropical area of a size similar to that of the Galápagos could be expected to have well over 100 bee species and 300 ant species but the Galápagos have only one bee and about 20 native ants.

Modes of dispersal to the Galápagos

Thus, insect representation at the family level is vastly different from that in the Neotropical fauna. The cause is the inequality of families in their ability to successfully complete both the sequential processes of dispersal and then colonization. Dispersal is a property of all the ecological, behavioral, and physiological characteristics of the taxon (the many species that comprise the family) and of mode and frequency of transport opportunity. Colonization is a property of both the life history requirements of the taxon and the characteristics of the new environment. I will first examine dispersal.

How do insects get to oceanic islands? It is evident that there are four main transport modes for potential island colonists: (1) actively or passively moving through the air by flying or being carried by the wind; (2) hitchhiking on (through either phoresy or ectoparasitism) or in (through endoparasitism) other organisms such as insects, mammals, reptiles, and birds; (3) floating or rafting or swimming in or on the ocean; and (4) intentional or accidental introduction by humans. The modes are of differing importance to different groups of Galápagos insects (Table 3.1) and for different islands or island groups worldwide.

1. Aerial (active flight and/or passive wind) transport. This probably accounts for well over 50% of the terrestrial arthropods of the Galápagos (Table 3.1). Many spiders must have arrived by ballooning of the juveniles on silk strands. Most small insects have undoubtedly been blown by winds as aerial plankton and about 40% of the beetles and most of the smaller lepidopterans and other small-sized winged insects may have arrived by such passive aerial transport. Some insects, such as dragonflies, macro-lepidoptera (larger moths and butterflies, which together are about 45% of the Lepidoptera fauna), and *Schistocera* locusts are known to be strong long-distance fliers and probably arrived by random, but wind-directed, flight. The mean body size of the Galápagos insect fauna appears to me to be smaller than for a mainland Ecuadorian fauna (but measurements are available for neither). Darwin first noted the small size of the insect fauna (see the Prologue). The smaller mean body size

Table 3.1. Estimates of relative importance for each insect order of the modes of natural dispersal of original colonist organisms to the Galápagos Archipelago; as percent of the total number of species in the order.

Order	Number colonizing species	Percent Aerial	Percent on sea surface	Percent phoreitc or parasitic
Collembola	33	0	100	0
Diplura	1	0	100	0
Archeognatha	1	0	100	0
Thysanura	2	0	100	0
Odonata	8	100	0	0
Orthoptera	16	25 ?	75 ?	0
Mantodea	1	0	100	0
Blattodea	3–4	0	100	0
Isoptera	4	0	100	0
Dermaptera	2	0	100	0
Embidina	1	0	0	100
Zoraptera	1	0	100	0
Psocoptera	20	100	0 ?	0
Thysanoptera	49	100	0	0
Phthiraptera	80	0	0	100
Hemiptera	83	75 ?	25 ?	0
Homoptera	50	75 ?	25 ?	0
Neuroptera	8	100	0	0
Coleoptera	313 ?	39	60	<1
Strepsiptera	1	0	0	100
Lepidoptera	250+	100	0	0
Diptera	200 ?	94	0	6
Siphonoptera	1	0	0	100
Hymenoptera	249 ?	95 ?	0	5 ?
Total	1375			

would support the idea that the majority of the insect colonizers were small and carried by winds. In contrast to its importance for insects, it may seem surprising that wind transport may account for only 9% of natural seedplant colonizations of the Galápagos (Porter 1976).

I have experimentally attempted to measure the importance of aerial transport of Galápagos insects. Nets were suspended from boat masts while traveling between the islands. In 1992, during a strong El Niño, with very wet conditions, these nets recovered 18 250 specimens of insects and spiders from a volume of 690 000 m^3 of air (Peck 1994a). The catch (summarized in Table 3.2) was very diverse and (as predicted) was of small body size. However, in 1996, similar net sampling of a volume of 1 130 000 m^3 of air yielded only 13 insects (Table 3.3) during a normal El Niño. Obviously, the seasonal and yearly abundance of insects in the air is correlated with the strength of the El Niño weather conditions. But over geological time, there has been and still is a significant probability for winged insects to move from continental sources to the Galápagos. Movement between the islands is probably most common during strong El Niño events. Because of this high potential for inter-island movement, we should expect very few winged insect species to be endemic to a single-island. This is, indeed, the observed general pattern.

In some groups of species, which are known to be mostly short-winged on the continent, our Galápagos collections contained mostly long-winged forms. Many phlaeothripid Thysanoptera (thrips) are polymorphic for wing length, and the short-winged or wingless adult morphs are the most common ones on continents. Selective dispersal by long-winged forms seems to be a likely explanation for their abundance on islands. Mound and O'Neill (1974: 497) found this in *Merothrips floridensis* Morgan and reported that island collected specimens were predominantly fully winged.

Linsley Gressitt and colleagues of the Bishop Museum once conducted a massive program to document the natural aerial transport of insects over the Pacific Ocean and elsewhere (e.g., Gressitt and Yoshimoto 1963; Holzapfel 1978; Holzapfel and Harrell 1968; Holzapfel et al. 1978). My study found 23 of the 101 identified orders and families of Gressitt and Yoshimoto (1963) and 8 families not in that list. The smaller air sample volumes and different faunal sources can reasonably account for the differences (Peck 1994*a*). But all these and other such studies demonstrate the immense numbers of insects that can move through the air, often over very long distances, especially during favorable climatic conditions.

2. Marine transport. A significant component of the total insect fauna probably arrived on the sea surface (Table 3.1), either on rafts of vegetation or flotsam or by floating themselves (as pleuston). This may be the most important mode for most of the flightless terrestrial arthropods. Schatz (1998) reports on 18 species of Oribatida mites that I recovered directly from the sea surface around the Galápagos. Estimates for the beetles suggest that marine transport may account for about 60% of the original colonists (Peck and Kukalova-Peck 1990), but a recalculation will be needed when the beetle fauna is fully itemized. Desender et al. (1992) discuss two species of carabid beetles that seem to have arrived in the strong El Niño of 1982–1983, perhaps by rafting. Flightless or poorly flying groups of large-bodied insects such as mantids, weevils, darkling beetles, and some (or all) families of crickets probably used this mode, as did millipeds, centipeds, terrestrial isopods, oribatida mites, and others. Termites and various

Table 3.2. Numbers of individuals of major groups of terrestrial arthropods recovered with nets from the air (690 000 m^3) or sea surface (130 000 m^2) between islands of the Galápagos during the strong El Niño of 1992 (from Peck 1994a, 1994c).The data show that large numbers of insects and other arthropods were in the air and on the sea between the islands at this time.

Taxon	Aerial	Sea Surface
Araneae	8	
Acari	0	150
Collembola	0	10
Thysanoptera	3	2
Orthoptera	0	6
Psocoptera	5	1
Homoptera		
Aleyrodidae	639	1
Psyllidae	184	2
Aphididae	17	87
Cicadellidae	8	56
Delphacidae	3	12
Hemiptera		
Cydnidae	0	14
Miridae	0	31
Nabidae	0	11
Others	1	10
Lepidoptera		
Noctuidae	0	8
Others	7	6
Coleoptera		
Chrysomelidae	2	0
Carabidae	0	16
Ptiliidae	0	27
Staphylinidae	0	15
Others	0	4
Diptera		
Tipulidae	10	34
Sciaridae	16 523	2778
Cecidomyidae	20	1
Ceratopogonidae	275	2
Chironomidae	17	44
Others	13	5

Table 3.2 (*concluded*).

Taxon	Aerial	Sea Surface
Hymenoptera		
Encyrtidae	394	8
Formicidae	110	470
Vespidae	2	13
Others	12	11
Totals	18 253	3 835

Table 3.3. Samples taken with 1 m^2 aerial net during boat travel between islands of the Galápagos Archipelago in the normal El Niño of 1996. These catches are very much lower than those from 1992, but are probably reflective of normal El Niño conditions, when there is comparatively little movement of insects between the islands.

Date	Between Islands	Time (hr)	Duration (hr)	Distance (km)	No. Insects
21.III	S. Cristobal to Española	1830–2130	3.5	50	6
21.–24.III	Española to S. Fe	2100–0100	5	68	0
24.III	S. Fe to S. Cruz	0800–1030	2.5	30	0
25.III	S. Cruz to Floreana	1845–2200	3.25	45	0
26.III	Gardner to Floreana	1030–1230	2.5	35	0
29.III	Floreana to S. Cruz	0830–1330	5	50	0
1–2.IV	S. Cruz to V. Alcedo	2100–0330	6.5	95	0
4.IV	V. Alcedo to Villamil	1630–2400	5.5	100	1
9.IV	Hermanas to S. Cruz	0930–1500	5.5	50	3
15.IV	S. Cruz to Campeón	1400–1730	3.5	55	0
23.IV	Floreana to Tortuga	0845–1200	3.25	52	0
10.V	S. Cruz to Isla Couley	0800–1330	5.5	100	0
10.V	Pt. Garcia to Isla Wolf	2100–1030	13.5	195	0
11.V	Wolf to Darwin	0830–1200	3.5	35	0
13.V	Darwin to Wolf	0915–1245	3.5	35	0
13–14.V	Wolf-Pta. Albemarle	1806–0600	10	135	3
Totals			82 h	1130 km	13 insects

other wood-boring insects probably arrived by rafting in wood as adults or immatures. All four species of Galápagos termites nest in or near wood, and all are present in the littoral zone and are thus salt-tolerant, so rafting can easily account for their presence. Crickets, mantids, and some of the other above examples may have colonized either as adults, immatures, or eggs in debris. Flightless *Gerstaeckeria* weevils probably arrived on rafting pieces of their *Opuntia* cactus host plants.

Several large-bodied wingless beetles such as endemic *Galapaganus* weevils and the three genera of Darwin's darklings (tenebrionids) are represented by species that occur on more than one island. Such cases are usually within the older eastern and central group of islands. It is logical that these species originated (speciated) on one island and that they have then moved from this to another island, probably after being washed to sea during heavy El Niño rainstorms and floods. Schatz (1998) observed that the endemic species of Oribatida mites that occur on more than one island are also most often distributed within the central or core islands of the Archipelago. Floating or rafting to and between the islands has obviously been important in the distribution process of most of the flightless species of arthropods, and most of these have been passively carried by oceanic currents. Only the *Halobates* sea-skaters (Gerridae) have arrived by active swimming on the sea surface.

The indigenous and endemic soil fauna is surprising in its diversity. Such small and physiologically fragile arthropods are usually thought to be poor "trans-oceanic" dispersers. While at least 16 eyeless soil arthropods are probably adventive and were accidentally caried by humans, at least another 10 are endemic, eyeless, wingless, depigmented species derived from naturally dispersing ancestral South American stocks with these same characteristics. To account for their presence in the Galápagos, it seems most likely that they arrived in an eyeless and wingless condition by natural rafting in soil (Peck 1990).

Various studies around the world (see Peck 1994*c*) have found live or intact insects of many orders floating unaided on the sea surface far from land. They had either been washed out to sea (both winged and wingless species) or had been blown to sea (winged species) and then fell to the sea surface. To experimentally study this phenomenon, we made pleuston net samples from 130 000 m^2 of sea surface in 1992, far at sea and between the islands of the Galápagos. We recovered 3841 specimens of live or recently dead acari and insects (Peck 1994*c*). The fauna was mostly all of winged species (Table 3.2), so these probably were blown out to sea and then fell to the sea surface.

It does seem that some insects can live for some time while floating unaided on the sea surface, but few experiments have studied this ability in flightless Galápagos insects. Table 3.4 contains data from some simple experiments, which seemingly demonstrate a low survival ability of some flightless Galápagos beetles floating on sea water.

Beaches could be examined immediately after storms as a way to measure what insect taxa have air-sea arrival potential (Howden 1977). Even with much searching I have never seen a natural raft of debris in Galápagos waters, but the presence of many logs and stumps from the continental mainland on Galápagos beaches shows that such "rafts" really do occur. It is estimated that floating

Table 3.4. Percent survival at 24, 48, and 60 h of flightless beetles on surface of seawater held in cups in shade at CDRS. The beetles floated well at the start, but lost much buoyancy after 24 h. This suggests that unaided surface floatation is probably not common or frequent as a dispersal mechanism for these large beetles.

Beetle	Sample Size	24 h	48 h	60 h
Ammophorus species	25	40%	16%	0%
Stomion species	35	34%	6%	0%
Galapaganus species	23	22%	9%	0%

materials can be transported from an Ecuadorian source such as the Gulf of Guayaquil in about two weeks (Thornton 1971). I suspect that the arrival on an island in the Galápagos of live sea-borne propagules is not very frequent in human time, and that successful colonization after arrival is an even rarer event. The endemic Galápagos tortoises, lizards, snakes, and rice rats are vertebrates that must have arrived by sea surface transport. There has been some past resistance to the idea that large vertebrates can be transported on the sea surface, but clear evidence is now available that this has occurred with West Indian iguanid lizards (Censky et al. 1998). No one has yet attempted to estimate or to measure the actual rate of sea-borne immigration of insects. At present, intense predation pressure in the littoral zone of the Galápagos by birds and lizards upon any insects arriving by sea would seem to seriously depress their establishment success, especially for medium- to large-bodied insects.

Plate II. A–B. Sampling aerial and sea surface insects between islands of the Galápagos. (A) Fishing boat with 1 m² aerial net fixed to mast to sample air-borne insects. (B) One m wide floating pleuston net towed at side of boat to sample sea surface insects.

Hence, both air and sea transport combined have been the most important modes of natural insect arrival. And these together are probably most frequent during strong El Niño climatic events when heavy mainland rains pour a flush of insects into continental air masses or rivers for wind and oceanic transport to the Archipelago. These same heavy rains stimulate insect emergence and movement by sea and air between the islands themselves. Through evolutionary time, such climatic events favoring dispersal may have occured with a rough average frequency of every 7 ± 4 years.

3. Transport on or in other animals. Insect ectoparasites, such as all of the 61 species of Phthiraptera (chewing bird-lice; = Mallophaga) which are reported in the literature (Linsley and Usinger 1966) and the 8 species of Hippoboscidae (louse flies; Diptera), as well as bird ticks, reptile ticks and chigger mites undoubtedly arrived on their vertebrate hosts. Bird transport has been important for seedplants, because it is estimated that 79% of the angiosperms arrived as propagules with birds, either on or in their feathers or in their digestive tracts (Porter 1976). Host birds have carried their bird-lice to the islands. The *Philornis* (Muscidae) flies, which feed as fly larvae on nestlings of Darwin's finches, but do not attach to them, probably colonized as adult flies independantly from their bird hosts. The two species of *Lasiurus* bats have carried their nycteribiid (bat fly; Diptera) ectoparasites. Marine mammals and rafting terrestrial mammals seem to have escaped some of their possible diversity of insect ectoparasites on the Galápagos. Fleas and sucking mammalian lice (= Anoplura) are conspicuously underrepresented, with only one and two indigenous or endemic species respectively. Reptiles have not entirely escaped their ticks and chiggers. A total of three species of trombiculid mites and four argasid and seven ixodid ticks are known on the reptiles (Schatz 1991). The conclusions from this are that birds, bats, and reptiles and their insect, tick and mite parasites could reach the islands much more easily than could non-flying land and marine mammals and their respective insect, tick and mite parasites.

Invertebrates themselves have also carried some of their own arthropod parasites (Table 3.1). One strepsipteran (in leafhoppers), one meloid blister beetle (on *Xylocopa* bees), one rhipiphorid beetle (in wood-boring beetle larvae), and several dryinid wasps and pipunculid flies (in leafhoppers) probably arrived as parasitic immatures on or in their immature and adult host insects. Parasitoid braconid, ichneumonid, and other abundant micro-hymenoptera wasps, as well as endoparasitic tachinid flies, probably colonized as adult insects, and not as larvae in their parasitized immature insect hosts.

A large number of scavenging and predatory insects occur in bird nests on continents (Hicks 1959) and certainly some Psocoptera (bark lice) and some other orders are phoretic (non-parasitic "hitchhikers") in bird feathers (Mockford 1967). It has been suggested that the endemic embidiid (web-spinner) was a bird-carried colonization (Ross 1966). In 1992 Ricardo Palma and I made a careful examination of 590 individual live land and sea bird hosts (73 species) for their phoretic and ectoparasitic arthropods (Palma and Peck, unpublished). We found a total of about 80 species of chewing bird lice plus a diversity of ectoparasitic mites, ticks, and flies (Hippoboscidae) totalling 3500 specimens. No non-parasitic arthropods were found on the birds. We concluded that, while

Plate III. A–F. Insect fieldwork in the Galápagos. (A) Base camp in the littoral zone of Isla Pinta. (B) Searching for ectoparasites in the colony of waved albatross on Isla Española. (C) Taking a break near the summit of Alcedo Volcano, Isla Isabela. (D) Collecting in the littoral zone on Isla Santa Fé. (E) Malaise trap in the humid forest zone of Isla Santiago. (F) Malaise trap in the arid zone of Isla Santa Fé in a dry year.

birds may theoretically carry phoretic arthropods to and between islands, it is a relatively rare event and has been relatively unimportant as a dispersal mechanism for free-living insects when compared to ectoparasitic species. On the other hand, wading birds seem to be the most likely candidates for the dispersal of eggs or propagules of some aquatic crustaceans, both on and in their bodies, from the mainland to and between islands with brackish coastal lagoons or temporary freshwater ponds (Peck 1994b).

No one has yet conducted a detailed study of the possible insect inhabitants of Galápagos bird nests. My casual observations suggest that there are very few insect inhabitants in the nests of Darwin's finches and the other land birds. The seaweed nests of the flightless cormorants mostly contain abundant fly maggots feeding on the rotting seaweed, and predatory staphylinid beetles feeding on these maggots. I suggest that sea birds which live in burrows, such as the Hawaiian petrel and Audubon's shearwater, will have the richest nest fauna, including more species of fleas. In my casual examination of inactive nests of gulls, boobys, frigatebirds, and pelicans, I have not found any insects.

4. Humans have intentionally brought many domestic animals and agricultural or horticultural plants to the Galápagos. Some of these have escaped and become feral. In the plants, 475 species are presently known to have been introduced and most of these were intentionally brought to the islands. At least 500 species are probably present. Unintentional introductions include rats, mice, and many weedy plants. I know of no examples of the intentional introduction of arthropods (even of honey bees), but there are at least 295 examples of unintentional introductions of adventive insect species (Peck et al. 1998) (Tables 1.4 and 3.5).

The first adventive insect may have arrived with the first European landings by Bishop Tomas de Berlanga and his party in 1535, as dermestid and *Necrobia* (clerid) beetles and cockroaches. These were all commonly associated with humans and stored products in their sailing ships. Pirates, who used the islands from shortly after the time of their discovery until the early 1700's, and whalers and sealers, from the mid 1700's to mid 1800's, may have brought other insects and an alleculid beetle (and other dry-wood insects such as bark-beetles) in logs or firewood from the mainland. The orders with the largest number of adventive species are Coleoptera, Homoptera, Lepidoptera, and Diptera. Some 60 beetles (mostly living in stored products or construction materials), many ants and 11 species of cockroaches are among the more commonly encountered species of insects introduced to date. Not all of the adventive species seem to have become permanently established. Some long-horned beeetles and the human bedbug have not been found since the original collection. The adventive species occur in greatest diversity on the four large islands with permanent human inhabitants.

Table 3.5. Summary of numbers of insect species probably adventive to the Galápagos, as categorized by their most likely modes of human transport.

On living plants	139
In dry stored agricultural and food products	42
In plant debris and soil around plants	37
Parasitoids in adventive insect hosts	20
In fresh or spoiled fruits and vegetables and food products	19
In dry goods, dunnage, pallets and packaging materials	12
In dry wood, construction materials and logs	10
On ships in general	6
On livestock and pets	4
Unknown	3
On rats (sucking louse)	1
Human ectoparasites, probably in bedding or clothing (human bedbug)	1
In water containers (*Culex* mosquito larvae?)	1
Total	295 species

Most of the adventive species (Table 3.5) are plant feeding insects such as thrips, aphids, small moths, leafhoppers, and scale insects. Exactly how many of these really occur yet remains to be taxonomically determined. Some of these phytophagous insects probably carried their predators or wasp parasitoids with them. Soil on plant roots carried earthworms, millipeds, centipeds, terrestrial isopods, silverfish, symphylans, schizomids, some beetles, some ants, and even such unlikely animals as a terrestrial flatworm and an onychophoran.

The little red fire ant, *Wasmannia auropunctata*, is the most disastrous arthropod introduction to date. It arrived early in the 1900's. It seems to have had a very serious impact on populations of endemic and indigenous arthropods (Lubin 1984). Other more recent and obviously deleterious introductions are as follows: (1). *Polistes versicolour* wasps now occur on all but the northern tier of islands. These are voracious predators on insect larvae such as caterpillars, which are also a food source for many endemic birds. (2). *Simulium bipunctatum* blackflies can now be a severe nuisance to humans in the highlands of Isla San Cristóbal. (3). *Brachygastra lecheguana* is a newly arrived stinging vespid wasp on Santa Cruz. (4). *Icerya purchasi*, the cottony cushion scale, is heavily infesting and maybe killing some trees on some islands, especially on Santa Cruz.

The rate of accumulation of adventive insect species is increasing, as are the numbers of permanent human inhabitants on the islands and visits by tourists (Fig. 3.1). I suggest that there is a cause and effect relationship. But, even with

Fig. 3.1. Summary by decade of the cumulative discovery of first records of insects adventive to the Galápagos Islands, compared to growth of the permanent human population and tourism visits (scaled to one-tenth). The rise in adventive insect species is strongly correlated with increases in both human settlement and tourism. Population data from Ecuador National Census and Galápagos National Park Services (MacFarland and Cifuentes 1996). After Peck et al. (1998).

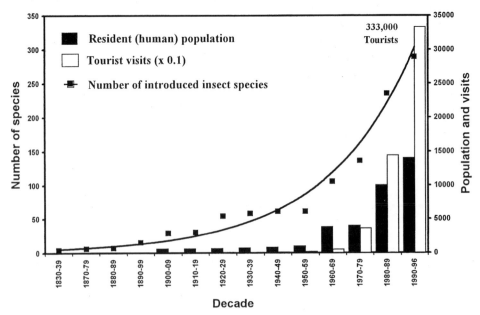

strict quarantine and control of incoming goods and people, the increase in commerce and movement of goods from the continent will probably see an unfortunate increase in the numbers of adventive insects in the Galápagos. This is what has happened even on islands with good quarantine procedures, such as Guam and Hawaii. Additionally, the increased numbers of tour ships moving between the islands of the Galápagos will continue to both dilute the indigenous and endemic faunas with these adventives (Silberglied 1978), and to mix the native faunas of the separate islands.

In summary, most insects have winged adults and these probably colonized the Galápagos through the air. The second most important transport mode has been on the sea surface. This accounts for flightless terrestrial arthropods and wingless free-living insects. The natural mode of least importance is transport on other animals, and is mostly important for ectoparasites and a few endoparasites. Humans have accidentally carried over 16% of the presently known insect fauna, and this proportion will probably continue to grow, even with increased quarantine procedures. It is difficult to predict how, when, or why the insect introductions will change the native plant and animal communities. Most have seemingly had a neutral environmental impact. Only a few have so far been serious.

Episodic or "punctuated" colonization

Present climatic conditions seemingly favor natural colonization during episodes of strong El Niño events. These have probably been recurring in the Galápagos throughout at least the last half of the recent (present) global *interglacial*. In contrast, during past global *glacial* periods, climatic events probably did *not* appreciably favor either the dispersal or colonization of insects. The reasons are complex and are as follows. In the last glacial period the Galápagos and western Ecuadorian Andes are known to have been drier than in the present interglacial period (Liu and Colinvaux 1985; Colinvaux et al. 1988). An important unanswered question is: what were the general glacial period *patterns* of wind and water circulation around the islands? There are three possibilities which have been suggested but for which there are no data of which I am aware. During glacials, the Intertropical Convergence Zone (ITCZ) (1) may not have moved south enough to bring annual precipitation (Colinvaux 1972), or (2) may have remained south of the archipelago (Newell 1973), or (3) oceanic currents and not the ITCZ may have changed (Houvenaghel 1974; Simpson 1975). Whatever the cause, the more stable climatic system of glacial times, with permanent inversions and little rain, probably made these times less suitable for the transport and establishment of insect colonists than during interglacial times. Thus, colonization was most likely episodic and occurred during strong El Niño climatic events of interglacials and was either rare or absent in glacial times.

Sources of the colonists

There is little detailed data on the mainland distributions of either indigenous Galápagos insects or mainland sister species of the endemic species. The general data now available seem to indicate that only a few of the Galápagos colonist insects came from southern South America (arid coastal Peru or Chile). Most of the faunal relationships are with the lowland semi-arid and seasonal Neotropics, along the Pacific coast of Central and South America, from Mexico to Ecuador. Only a few groups are known to have species also occurring on Cocos Island (Bickel and Sinclair 1997; Hogue and Miller 1981). The best phylogenetic and biogeographic analyses to date are for some mirid bugs (Schuh and Schwartz 1985), for 17 species of aleocharine rove beetles (Klimaszewski and Peck 1998), for 11 species of ceratopogonid midges (Borkent 1991), 15 species of ephydrid shore flies (Mathis 1995), 17 species of dolichopodid flies (Bickel and Sinclair 1997), and a clade of 15 species of *Galapaganus* weevils (Lanteri 1992). These all show a general biogeographic pattern of a western Neotropical source area and that the Galápagos species are relatively recently derived species. An exception is a clade of six sphaerocerid flies (Marshall 1997) which shows the Galápagos species to be *basal* (ancestral) to four others in the Neotropics.

I have independantly tried to evaluate the available evidence about the source areas of the insects. As a sample I took 130 indigenous insect species in all orders (except Diptera, Lepidoptera, Hymenoptera which are still too poorly known) for which mainland distributional data are available. I found that 65% are widespread Neotropical species, and most of the others represent subsets of areas in the Neotropics. These subsets are 17% with distribututions in Mexico and Central America, 4% from Venezuela, Colombia, or northern Ecuador, 4% from Ecuador, 6% from arid or semi-arid coastal Ecuador, Peru, Chile, or arid Argentina, and 1% are known only from islands of the West Indies. I think that it is reasonable to assume that the relatively recent colonization by the remaining Galápagos indigenous insects may have been from these areas in similar proportions, and that the source areas for the even earlier colonizations (for the now endemic species) may also have been similar. However, the West Indies seems to me not to be a real source area, because I interpret this pattern as an artifact of inadequate knowledge of mainland Neotropical insects. The remaining 3% are a bit of a surprise because they are seemingly not from anywhere in the Americas. They have a relationship with the central or western Pacific (Austral-Asian Region) and not the Neotropics.

The western Pacific seems to be a real source area for several species of terrestrial insects; a stenocephalid (*Dicranocephalus*) true bug (Froeschner 1985), a cantharid beetle (Wittmer and Peck 1993), an endomychid beetle, and some Psocoptera (Thornton and Woo 1973). One species of dolichopodid fly (*Thinophilus hardyi*) also occurs in the Hawaiian Islands, and has no close species in North or Central America (Bickel and Sinclair 1997). Additionally, two land snails (Chambers 1991) and 15 species of Oribatida mites (Schatz 1998) show a west or south Pacific connection. All these then have seemingly come from the west by bird, air or ocean transport (probably by the east flowing equatorial or north equatorial counter currents). These oceanic currents have certainly been important in bringing a significant marine faunal component of molluscs, crustaceans, and fishes from the Indo-West Pacific marine province (James 1991). And the above examples show the reality of this pattern for a small part of the terrestrial invertebrate fauna.

There is general agreement for other plant and animal groups that the principal source area of colonists was Mexico, or Central America or northwestern South America. This has been long appreciated for plants (Svenson 1946), and has been well analyzed for the angiosperms (Porter 1976); almost all (99%) of the indigenous species of vascular plants also occur in South America (Porter 1984).

Establishment of colonists

The earliest insect colonists encountered a harsh and seasonally arid climate, new and unvegetated barren lava, and a coastline with only basalt sand or cobble beaches (with no limey coral-algal sand). They were probably hardy, weed-like eurytopic species (with wide ecological tolerances). The base of the original food chain was supported only by marine drift and the bodies of sea animals. There were no wrack lines of marine algae or angiosperms such as turtle grass. The marine-detritus guild of vagile littoral zone scavengers (isopods, earwigs, various beach flies, some beetles) were probably the first terrestrial arthropods to establish, and were followed by generalist predators such as centipeds and carabid and staphylinid beetles. The establishment of sea bird colonies could have preceded land colonization by plants. Sea birds pour a heightened amount of energy, as guano, from marine into terrestrial communities. This makes possible the establishment of a more diverse array of terrestrial arthropods (Polis and Hurd 1996). The first vascular plants could have been mangroves, *Ipomoea* beach morning glories, and other littoral zone plants with good dispersal potential. Insect faunas of these strand plants may have the ability to track their hosts, but such island insects seem to be poorly studied. Plants and insects with strict reproductive interdependence are unlikely to arrive together, so any such colonists would have perished.

After arrival, potential colonists must survive and reproduce. A male and female, or an inseminated female, or a self-fertilizing (parthenogenetic) individual is the minimum requirement for reproduction. In the source area the most successful candidates for colonization would have the properties of being abundant, polyphagous, with high genetic variability, short generation time, and able to withstand a wide variety of physical conditions. These are the characteristics usually associated with an "r" reproductive strategy, where maximum reproductive capacity is selected or favored. Parthenogenesis itself is actually or potentially known in at least 8 Galápagos insect species or groups of species and these are all adventive species (Table 3.6). So, while parthenogenesis has seemingly not aided natural insect colonization in the Galápagos, it has probably been important in the Galápagos Oribatoid mites, especially Lohmaniidae (Schatz 1998).

The colonization of new and young volcanic landscapes is probably not a frequent event. There is a lack of food sources from the primary production of plants, and they are physiologically harsh in their aridity and high daytime temperatures. Finely divided volcanic ash can abrade the cuticular lining of insect respiratory and digestive tracts or abrade the cuticular waterproof body covering, or it may have physical sorbtive qualities. Water loss (especially in juveniles) is thus increased and survival lowered (Alexander et al. 1944; Edwards 1986; Edwards and Schwartz 1981; 1945, 1947). Brown and bin Hussain (1981)

Table 3.6. Insects in the Galápagos which are probably parthenogenetic. Parthenogenesis allows for island colonization by a single unfertilized female. While this may be important for other archipelagos, all Galápagos examples seem to have arrived by human introduction, and not by natural means.

Taxon	Mode of arrival
Thysanura	
Nicoletia meinerti (Nicoletiidae)	adventive
Blattodea	
Pycnoscelus surinamensis (Blaberidae)	adventive
Psocoptera	
Lachesilla aethiopica (?) (Lachesillidae)	adventive
Peripsocus pauliana (?) (Peripsocidae)	adventive
Archipsocus spinosus (Archipsocidae)	adventive
Homoptera	
Aphids (many are facultatively parthenogenetic)	adventive
Scale insects? (many are facultatively parthenogenetic)	adventive
Hymenoptera	
Venturia canescens (Ichneumonidae)	adventive

found that active predaceous and parasitic insects were particularly susceptible to volcanic ash after the 1980 Mt. St. Helens erruption. Volume 37 of the journal Melanderia contains other papers on the impact of volcanic ash from this eruption on insects.

Allobiosphere habitats. Barren habitats which lack their own photosynthetic base for a food chain are sometimes called the "allobiosphere." But even these may have communities of predatory and scavenging insects. Their food comes by importation from adjacent productive habitats. Such communities (of spiders, beetles, crickets, etc.) are known in barren non-productive habitats such as ash fields, new lava flows, and lava tube caves in both the Hawaiian and Canary archipelagos (Howarth 1987), and on the volcanic Anak Krakatau island of Indonesia (Thornton 1991, 1996). These allobiosphere-specialists must be good dispersers if they are to locate such ecologically ephemeral habitats. And they are observed to vanish when plant succession begins on an ash field or a lava flow. The data seem to indicate that these specialist inhabitants require food materials carried as wind-blown fall-out from neighbouring terrestrial communities, or carried as sea-drift to lava flows at the seashore. The species themselves seem to have evolved from occupants of neighbouring communities, and they may not be good at long distance inter-island dispersal themselves, but exceptions exist (as in the *Speonemobius* crickets of Anak Krakatau, New and Thornton 1988). No evidence for the existence of such an arthropod allobiosphere community has yet been found in the Galápagos through the use of baited traps

(Peck 1996*a*) in young lava flows, but the search should continue. The use of pan traps on ash fields to measure the aerial input of food as fallout has not yet been attempted.

Trophic generalists. Colonization is probably easier for trophic generalists (scavengers and predators) than for herbivores which are more likely to be specialist feeders. Becker (1975) found that island beetle faunas in general tend not to be as rich in herbivores as the faunas on continents. In an analysis of the trophic structure of Galápagos beetles (Peck and Kukalova-Peck 1990) there were more trophic generalists (scavengers and predators) and fewer herbivores. However, in Heteroptera bugs, colonization of islands by herbivores seems to be more successful than by predators globally (Becker 1992). This is also true with new data for Galápagos Heteroptera, yielding 27 species of predators against 61 species of herbivores, in the same families selected by Becker. Why beetles and bugs have opposite proportions of herbivores is not apparent.

Trophic specialists. There is little evidence that Galápagos insects have narrrow or restricted feeding needs. The few examples are *Gerstaeckeria* weevils which can feed only on the tissues of *Opunti*a cactus, and the the widespread and migratory Monarch butterfly, *Danaus plexippus*, which feeds only on the milkweed *Asclepias curassavica,* a plant brought to the islands by humans. Monarchs were unreported in the Galápagos until first seen by William Beebe in 1924. It is unlikely that they would have escaped the notice of earlier collectors. There is another asclepiad milkweed species on the islands (*Sarcostemma angustissima*), but the Monarch is unable to develop on it. It seems likely that the presence of the Monarch is a natural colonization that could occur only after humans introduced its host plant (Roque 1998).

Ecological escape. Plant or animal colonists on islands may be ecologically "released" through escape from their continental herbivores, parasites, predators, and competitors. Escape from parasites is evident in some Galápagos vertebrates, but has not been recognized before in Galápagos insects. The host insect species must be already established at the time of arrival of their predators, parasites or parasitoids A diverse assemblage of nine families of aquatic insects in South America are usually hosts to the parasitic larvae of water mites. Since the mites are absent in the Galápagos, it is assumed that the insects in these families in the Galápagos have escaped these parasites (Gerecke et al. 1995). But any effect of the release from parasitism on the insect populations is not apparent.

Many cases of escape from insect herbivores or predators must exist, but few are recognized. One example is the seed-producing legume plants which have escaped many (but not all) of their seed predator bruchid beetles. The bruchid *Megacerus leucospilus* feeds on the seeds of the widespread beach morning-glory *Ipomoea pes-caprae* in Central America, but the plant seems not to have this seed predator on the Galápagos. The larvae of many phycitine pyralid moths feed on cacti in the Americas, but none of these are known from the cacti of the Galápagos (Landry and Neunzig 1997). While working with grasshoppers and crickets, D. Otte and I have not found them to be parasitized by tachinid fly larvae or nematomorph (horse-hair) worms, which are their common parasites on the continent (D. Otte, pers. comm). I would expect the parasitoid load on many Galápagos insects to be lower, at least in diversity if not in intensity, than in the mainland Neotropics, but this has not been studied.

Few insect pathogens are known in the Galápagos. Entomopathogenic fungi which attack scale insects are reported by Evans and Samson (1982). Another fungus has been reported from *Pheidole* ants on Santiago Island (Espadaler 1997).

References

Alexander, P., Kitchener, J.A., and Briscoe, H.V.A. 1944. Inert dust insecticides. I. Mechanisms of action. Ann. Appl. Biol. **31**: 143–149.

Becker, P. 1975. Island colonization by carnivorous and herbivorous Coleoptera. J. Anim. Ecol. **44**: 893–906.

Becker, P. 1992. Colonization of islands by carnivorous and herbivorous Heteroptera and Coleoptera: effects of island area, plant species richness, and 'extinction rates'. J. Biogeogr. **19**: 163–171.

Bickel, C.J., and Sinclair, B.J. 1997. The Dolichopodidae (Diptera) of the Galápagos Islands, with notes on the New World Fauna. Entomol. Scand. **28**: 241–270.

Borkent, A. 1991. The Ceratopogonidae (Diptera) of the Galápagos Islands, Ecuador with a discussion of their phylogenetic relationships and zoogeographic origins. Entomol. Scand. **22**: 97–122.

Brown, J.A., and bin Hussain, Y. 1981. Physiological effects of volcanic ash upon selected insects in the laboratory. Melanderia **37**: 30–38.

Censky, E.J., Hodge, K., and Dudley, J. 1998. Over-water dispersal of lizards due to huricanes. Nature **395**: 556.

Chambers, S.M. 1991. Biogeography of Galápagos land snails. *In* Galápagos marine invertebrates: taxonomy, biogeography and evolution in Darwin's Islands. *Edited by* M.J. James. Topics in Geobiology, vol. 8. Plenum Press, New York. pp. 307–325.

Colinvaux, P.A. 1972. Climate and the Galápagos Islands. Nature **240**: 17–20.

Colinvaux, P.A., Frost, M., Frost, I., Liu, K.B., and Steinitz-Kraven, M. 1988. Three pollen diagrams of forest disturbances in the western Amazon Basin. Rev. Paleobot. Palynol. **55**: 73–81.

Desender, K., Baert, L., and Maelfait, J.-P. 1992. El Niño events and establishment of ground beetles in the Galápagos Archipelago. Bull. Inst. R. Sci. Nat. Entomol. Belg. **62**: 67–74.

Edwards, J.S. 1986. Arthropods as pioneers: recolonization of the blast zone on Mt. St. Helens. Northwest Environ. J. **2**: 63–73.

Edwards, J.S, and Schwartz, L.M. 1981. Mount St. Helens ash: a natural insecticide. Can. J. Zool. **59**: 714–715.

Espadaler, X. 1997. *Pheidole williamsi* (Hymenoptera: Formicidae) parasitized by *Myrmiciniosporidium durum* (Fungi) on San Salvador island (Galápagos Islands). Sociobiology **30**: 99–101.

Evans, H.C., and Samson, R.A. 1982. Entomogenous fungi from the Galápagos Islands. Can. J. Bot. **60**. 2325–2333.

Froeschner, R. 1985. Synopsis of the Heteroptera or true bugs of the Galápagos Islands. Smith. Cont. Zool. **407**: 1–84.

Gerecke, R., Peck, S.B., and Pehofer, H.E. 1995. The invertebrate fauna of the inland waters of the Galápagos Archipelago (Ecuador); a limnological and zoogeographical summary. Arch. Hydrobiol. Suppl. **107**: 113–147.

Gressitt, J.L., and Yoshimoto, C.M. 1963. Dispersal of animals in the Pacific. *In* Pacific Basin biogeography. A symposium. *Edited by* J.L. Gressitt. Bishop Museum Press, Honolulu. pp. 283–292.

Hicks, E.A. 1959. Checklist and bibliography of the occurrence of insects in bird nests. Iowa State College Press, Ames, IA.

Hogue, C.L., and Miller, S.E. 1981. Entomofauna of Cocos Island, Costa Rica. Atoll Res. Bull. **250**: 29 pp + additions 4 pp.

Holzapfel, E.P. 1978. Transoceanic airplane sampling for organisms and particles. Pac. Insects **18**: 169–189.

Holzapfel, E.P., and Harrell, J.C. 1968. Transoceanic dispersal studies of insects. Pac. Insects **10**: 115–153.

Holzapfel, E.P., Clagg, H.B., and Goff, M.L. 1978. Trapping of air-borne insects on ships in the Pacific, Part 9. Pac. Insects **19**: 65–90.

Houvenaghel, G.T. 1974. Equatorial undercurrent and climate in the Galápagos Islands. Nature **250**: 565–566.

Howarth, F.G. 1987. Evolutionary ecology of aeolian and subterranean habitats in Hawaii. Trends Res. Ecol. Evol. **2**: 220–227.

Howden, H.F. 1977. Beetles, beach drift and island biogeography. Biotropica **9**: 53–57.

James, M.J. (*Editor*). 1991. Galápagos marine invertebrates: taxonomy, biogeography and evolution in Darwin's Islands. Topics in Geobiology, vol. 8. Plenum Press, New York.

Klimaszewski, J., and Peck, S.B. 1998. A review of Aleocharine rove beetles from the Galápagos Islands, Ecuador (Coleoptera: Staphylinidae: Aleocharinae). Rev. Suisse Zool. **105**: 1–40.

Landry, B., and Neunzig, H.H. 1997. A review of the Phycitinae of the Galápagos Islands (Lepidoptera: Pyralidae). Entomol. Scand. **28**: 493–508.

Lanteri, A.A. 1992. Systematics, cladistics, and biogeography of a new weevil genus, *Galapaganus* (Coleoptera: Curculionidae) from the Galápagos Islands, and coasts of Ecuador and Peru. Trans. Am. Entomol. Soc. **118**: 227–267.

Linsley, E.G., and Usinger, R.L. 1966. Insects of the Galápagos Islands. Proc. Cal. Acad. Sci. (4) **33**: 113–196.

Liu, K.-B., and Colinvaux, P.A. 1985. Forest changes in the Amazon Basin during the last glacial maximum. Nature **318**: 556–557.

Lubin, Y.D. 1984. Changes in the indigenous fauna of the Galápagos Islands following invasion by the little red fire ant, *Wasmannia auropunctata*. Biol. J. Linn. Soc. **21**: 229–242.

MacFarland, C., and Cifuentes, M. 1996. Case study: Ecuador. *In* Human population, biodiversity and protected areas: science policy issues. *Edited by* V. Dompka. Report of a workshop, April 20–21, 1995, Washington, DC. AAAS. International Directorate, Washington, DC. pp. 135–188.

Marshall, S.A. 1997. A revision of the *Sclerocoelus galapagensis* group (Diptera: Sphaeroceridae: Limosininae). Insecta Mundi **11**: 97–115.

Mathis, W.N. 1995. Shore flies of the Galápagos Islands (Diptera: Ephydridae). Ann. Entomol. Soc. Am. **88**: 627–640.

Mockford, E.L. 1967. Some Psocoptera from plumage of birds. Proc. Entomol. Soc. Wash. **69**: 307–309.

Moreno Espinosa, M., Silva del Pozo, X., Estévez Jácome, G., Marggraff, I., and Marggraff, P. 1997. Mariposas del Ecuador. Occidental Exploration and Production Company, Colección "El Ecuador Secreto." Imprenta Mariscal, Quito.

Mound, L.A., and O'Neill, K. 1974. Taxonomy of the Merothripidae with ecological and phylogenetic considerations (Thysanoptera). J. Nat. Hist. **8**: 481–509.

New, T.R., and Thorton, I.W.B. 1988. A pre-vegetation population of crickets subsisting on allochthonous aeolian debris on Anak Krakatau. Phil. Trans. R. Soc. Lond. B **322**: 481–485.

Newell, R.E. 1973. Climate and the Galápagos Islands. Nature **245**: 91–92.

Peck, S.B. 1990. Eyeless arthropods of the Galápagos Islands, Ecuador: composition and origin of the cryptozoic fauna of a young tropical oceanic archipelago. Biotropica **22**: 366–381.

Peck, S.B. 1992. The dragonflies and damselflies of the Galápagos Islands, Ecuador. Psyche **99**: 309–321.

Peck, S.B. 1994*a*. Aerial dispersal of insects between and to islands in the Galápagos Archipelago, Ecuador. Ann. Entomol. Soc. Am. **87**: 218–224.

Peck, S.B. 1994*b*. Diversity and zoogeography of the non-oceanic Crustacea of the Galápagos Islands, Ecuador (excluding terrestrial isopoda). Can. J. Zool. **72**: 54–69.

Peck, S.B. 1994*c*. Sea-surface (pleuston) transport of insects between islands in the Galápagos, Archipelago, Ecuador. Ann. Entomol. Soc. Am. **87**: 576–582.

Peck, S.B. 1996*a*. The arthropods of the allobiosphere (barren lava flows) of the Galápagos Islands, Ecuador. Noticias de Galápagos **56**: 9–12.

Peck, S.B. 1996*b*. Origin and development of an insect fauna on a remote archipelago: The Galápagos Islands, Ecuador. *In* The origin and evolution of Pacific Island biotas, New Guinea to eastern Polynesia: patterns and processes. *Edited by* A. Keast and S. Miller. SPB Academic Publishing, Amsterdam, The Netherlands. pp. 91–122.

Peck, S.B., and Kukalova-Peck, J. 1990. Origin and biogeography of the beetles (Coleoptera) of the Galápagos Archipelago, Ecuador. Can. J. Zool. **68**: 1617–1638.

Peck, S.B., Heraty, J., Landry, B., and Sinclair, B.J. 1998. The introduced insect fauna of an oceanic archipelago: The Galápagos Islands, Ecuador. Am. Entomol. **44**: 218–237.

Polis, G.A., and Hurd, S.D. 1996. Linking marine and terrestrial food webs: allochthonous input from the ocean supports high secondary productivity on small islands and coastal land communities. Am. Nat. **147**: 396–423.

Porter, D.M. 1976. Geography and dispersal of Galápagos vascular plants. Nature **264**: 243–252.

Porter, D.M. 1984. Relationships of the Galápagos flora. Biol. J. Linn. Soc. **21**: 243–251.

Roque, L. 1998. The Monarch butterfly in the Galápagos Islands: it it a native or an introduced species? Noticias de Galápagos **59**: 9–11.

Ross, E.A. 1966. A new species of Embioptera from the Galápagos Islands. Proc. Cal. Acad. Sci. 4 ser, **34**: 499–504.

Schatz, H. 1991. Catalogue of known species of Acari from the Galápagos Islands (Ecuador, Pacific Ocean). Int. J. Acarol. **17**: 213–225.

Schatz, H. 1998. Oribatid mites from the Galápagos Islands — faunistics, ecology and speciation. Exper. Appl. Acarol. **22**: 373–409.

Schuh, R.T., and Schwartz, M.D. 1985. Revision of the plant bug genus *Rhinacloa* Reuter with a phylogenetic analysis (Hemiptera: Miridae). Bull. Am. Mus. Nat. Hist. **179**: 379–470.

Silberglied, R. 1978. Inter-island transport of insects aboard ships in the Galápagos Islands. Biol. Conserv. **13**: 273–278.

Simpson, B.B. 1975. Glacial climates in the eastern tropical south Pacific. Nature **253**: 34–36.

Svenson, H.K. 1946. Vegetation of the coast of Ecuador and Peru and its relation to the Galápagos Islands. Am. J. Bot. **33**: 394–498.

Thornton, I.W.B. 1971. Darwin's Islands, a natural history of the Galápagos. Natural History Press, Garden City, NY. 322 pp.

Thornton, I.W.B. 1991. Krakatau — studies on the origin and development of a fauna. *In* The unity of evolutionary biology. Proceedings of the Fourth International Congress of Systematic and Evolutionary Biology. *Edited by* E.C. Dudley. Dioscorides Press, Portland, OR. pp. 396–408.

Thornton, I. 1996. Krakatau: the destruction and reassembly of an island ecosystem. Harvard University Press, Cambridge, MA.

Thornton, I.W.B., and Woo, A.K.T. 1973. Psocoptera of the Galápagos Islands. Pac. Insects **15**: 1–58.

Wigglesworth, V.B. 1945. Transpiration through the cuticle of insects. J. Exp. Biol. **21**: 97–114.

Wigglesworth, V.B. 1947. The site of action of insect dusts on certain beetles infesting stored products. Proc. R. Entomol. Soc. London **A22**: 65–69.

Wittmer, W., and Peck, S.B. 1993. Cantharidae and Malachiidae of the Galápagos Islands, Ecuador. Coleop. Bull. **47**: 77–81.

Chapter 4
Processes: evolution of the insect fauna

After the arrival and establishment of the earlier colonist insect species, they were usually geographically isolated from the continental source populations. Through this isolation the island populations could genetically diverge through time into new and endemic species. This is the process of allopatric speciation. Within the vertebrates, the most famous example of speciation is the development of a group of 13 descendant species of Darwin's finches from one ancestral colonization (Grant 1986). The famous Galápagos tortoises are an example of sub-specific diferentiation. But, in contrast with the many studies of the vertebrates, the results of the evolutionary differentiation of species and their ecological assemblages have rarely been considered for Galápagos insects.

Speciation

There are several insect genera which have undergone appreciable sub-speciation or speciation in the Galápagos (Table 4.1). But none of these have equalled the dramatic swarms of species of insects, snails, or birds that have evolved in Hawaii (Table 4.2). For instance, there is only minor speciation in Galápagos *Drosophila* (Carson et al. 1983), while hundreds of species occur in Hawaii. Generally, when the entire indigenous insect fauna of the Galápagos is considered, there has been comparatively little speciation within each colonizing genus (Tables 4.2 and 4.3), with a mean of 1.15 species per genus.

The insects grouped as adventive, indigenous, endemic, and paleo-endemic probably represent increasing age or times of arrival descending from modern, to near recent, to past, and distant past dispersal events and isolation processes. The adventive species have been carried to the islands in historic times. The indigenous species undoubtedly mostly evolved in the Neotropics and then naturally dispersed to the Galápagos, probably within the past several hundred thousand years. They have arrived so recently in evolutionary time that they have not yet differentiated in isolation, and may even still be in genetic contact with mainland populations through continued dispersal. But much of the Galápagos insect fauna (from 1/3 to 2/3 depending on order) is composed of endemic species (Table 1.5). For these, their ancestral colonization was enough in the geological past that differentiation through geographic and genetic isolation has occurred. Of the over 1000 insect genera which are present, only 24 are thought to be endemic (Table 4.4) and these must represent very early colonizations and can be considered as among the oldest colonists, or as paleo-endemics.

Table 4.1. Some notable examples of endemic species groups or species swarms in the terrestrial invertebrate biota of the Galápagos. Single colonizations remain to be demonstrated by cladistic or other methods. In some cases flightlessness evolved after colonization. The ± symbol indicates that only some of the species in the genus in the Galápagos are now flightless.

	Number of species or subspecies	Number of ancestral colonizations	Now flightless?
Mollusca; Gastropods			
Bulimulus land snails	65	1	yes
Orthoptera			
Gryllus (Gryllidae)	8	2 ?	±
Halmenus (Acrididae)	4	1	yes
Conocephalus (Tettigoniidae)	3	2 ?	±
Pteronemobius (Gryllidae)	2	1	yes
Blattodea			
Ischnoptera (Blattellidae)	3	1	yes
Chorisoneura (Blattellidae)	2	1	±
Hemiptera			
Dagbertus (Miridae)	12	5	no
Ghinallelia (Reduviidae)	8	1	yes
Creontiades (Miridae)	7	4	no
Cyrtopeltis (Miridae)	7	1	no
Psallus (Miridae)	4	1	no
Jadera (Rhopalidae)	4	4	no
Nabis (Nabidae)	3	2	yes
Corythaica (Tingidae)	3	2	yes
Homoptera			
Philatis (Issidae)	20	1 ?	yes
Scaphytopius (Cicadellidae)	17	2 ?	no
Oliarus (Cixiidae)	12	1 ?	yes
Nesosydne (Delphacidae)	7	2 ?	yes
Coleoptera			
Pterostichus (Carabidae)	12	2 ?	yes
Blapstinus (Tenebrionidae)	16	3 ?	±
Ammophorus (Tenebrionidae)	12	2 ?	yes
Stomion (Tenebrionidae)	12	1	yes
Galapaganus (Curculionidae)	6	2–3	yes
Pseudopentarthrum (Curculionidae)	6	1?	yes
Diptera			
Euxesta (Otitidae)	8	1 ?	no
Nocticanace (Canaceidae)	8	2 ?	no
Dasyhelea (Ceratopogonidae)	7	7	no
Hymenoptera			
Enicospilus (Ichneumonidae)	6	4	no

Table 4.2. Comparisons of total numbers of native species and mean numbers of species descending from an ancestral colonizing species on two archipelagos. Comparisons of equivalent groups are unfortunately not always possible. The differences between Hawaii and Galápagos result from the sum of physical properties such as island age and isolation, and biological properties such as habitat diversity, taxon vagility, and genetic characteristics. The recent numbers for Hawaii in Eldredge and Miller (1995, 1997) cannot be used for comparison because they do not provide an estimate of numbers of ancestral colonizing species.

Archipelago	Taxon	Number present species	Estimated number colonizing species	Mean number species per ancestor
Hawaii	flowering plants	1894	272	7
	reptiles	0	0	0
	land and shore birds	70	15	5
	land molluscs	1064	22	48
	all insects	5005	250	20
	beetles	1388	70	20
Galápagos				
	vascular plants	541	411	1.32
	reptiles	20	10	2
	land birds	57	43	1.33
	land molluscs	80	16?	5.0
	all insects	1558	1357	1.15
	beetles	358	313	1.14

Table 4.3. Means of species:genus ratios of the native taxa of the smaller insect orders of the Galápagos Islands. The small orders whose species:genus ratio does not exceed 1.0 are not listed.

Order	Native species	Native genera	Species: Genus ratio
Homoptera	85	28	3.04
Blattodea	7	3	2.33
Orthoptera	29	13	2.23
Hemiptera	113	58	1.95
Collembola	33	21	1.57
Isoptera	4	3	1.33
Odonata	8	7	1.14

Four genera, once thought to be Galpápagos endemics, have since been found in mainland Neotropical localities (Table 4.4) and others may yet be detected.

Colonization of the islands has usually been by only one species in a genus. Most genera are represented by only one (occasionally an endemic) species, and this pattern was first noted by Darwin (see Prologue). Only a minority of the

Table 4.4. Endemic Genera of Galápagos Insects (see also Table 4.1), and if the species are flightless. This shows that generic endemism is correlated with flightlessness only about half the time.

	Number of species	Flightless
Mantodea		
Galapagia	1	females only
Orthoptera		
Closteridea (Acrididae)	1	yes
Halmenus (Acrididae)	4	yes
Hemiptera		
Darwinysius (Lygaeidae)	2	yes
Galapagocoris (Miridae)	1	yes
Galapagomiris (Miridae)	1	yes
Undescribed genus (Saldidae*)*	1	yes
Trincavellius (Pentatomidae)	1	yes
Neuroptera		
Galapagoleon (Myrmeleontidae)	1	no
Coleoptera		
Acribus (Nitidulidae)	1	no
Blairinus (Anobiidae)	2	no
Docemus (Chrysomelidae)	1	no
Galapagodacnum (Hydrophilidae)	1	no
Mystroceridius (Carabidae)	2	yes
Neolinus (Staphylinidae)	1	no
Neoryctes (Scarabaeidae)	4	yes
Pinostygus (Staphylinidae)	1	yes
Stomion (Tenebrionidae)	11	yes
Diptera		
Galapagomyia (Sarcophagidae)	1	no
Galapagosia (Tachinidae)	1	no
Gigantotheca (Sarcophagidae)	3	no
Opsophytopsis (Sarcophagidae)	1	no
Paraeuxesta (Otitidae)	6	no
Sarothromyiops (Sarcophagidae)	1	no

Genera once thought to be endemic and which are now known to occur elsewhere

Hemiptera	also in:
Darwinivelia (Mesoveliidae)	Colombia, Brazil
Coleoptera	
Nesoeme (Cerambycidae)	Ecuador ?
Nesozineus (Cerambyridae)	Ecuador
Pantomorus (= *Galapaganus*) (Curculionidae)	Ecuador, Peru

genera which are present (about 50 out of over 1000) have experienced much species multiplication on the islands. The process of forming several species by allopatric speciation on a single individual island has not been a dominant evolutionary process in Galápagos insects, while it has been a spectacularly exuberant process in the Hawaiian Archipelago. This reflects, in comparison to Hawaii, that the Galápagos are younger, have had more frequent colonization from the mainland, are less isolated from mainland populations after colonization, and that there is more subsequent between-island colonization within the archipelago. In the more vagile insects (the winged species) this results in a lower level of between-island speciation and is easy to understand. Single island endemism is more prevalent in flightless arthropods of low vagility and these are usually restricted to either the arid lowlands or the moist uplands of a single high island. Schatz (1998) found patterns in Oribatida mites that suggest speciation on different islands, and that some multiple speciation has occurred on single islands by the occupation of different habitats. Insect examples of more speciation in less vagile groups (Table 4.1) are in carabid beetles, weevils, issid bugs, and Darwin's darkling beetles (*Stomion, Ammophorus*, and *Blapstinus*). Interestingly, even within flightless genera in the arid lowlands, many species occur on more than one island, and these are probably evidence of inter-island oceanic transport following the origin of the species on one island (Finston and Peck 1997).

Few groups of organisms present equivalent amounts of endemism. This is obviously a result of differences in their vagility and the amount of gene flow between continental and island populations. In insects, the good dispersers have lower levels of endemism, while poorer dispersers have higher levels (Tables 1.5, 4.1, 4.4). Comparison of the Galápagos and Hawaiian archipelagos shows a much greater mean number of speciation events from a single colonist ancestor in Hawaii (Table 4.2). This is probably the result of Hawaii's greater age, area, ecological diversity, and isolation (this is to say that colonist arrival is less frequent, and that genetic dilution of island populations by mainland genomes is also less frequent).

Loss of wings. Loss of flight ability is one of the more pronounced phenomena associated with island insects. This is seemingly not a property of island life itself, but of habit stability and homogeneity (Roff 1991). Flightlessness can also be seen to be correlated with evolution toward a "K" reproductive strategy, which is a lowering of reproductive capacity, but with higher survival rates for each individual offspring. This is also correlated with stable and homogeneous habitats. Perhaps, in light of the variable climatic history of the Galápagos, it is surprising that so much shift to "K" reproductive characteristics has occurred.

Flightlessness also frequently occurs in insects in semi-arid habitats, which is the best single characterization of Galápagos environments. Beetles are prime examples. Chown et al. (1998) have shown that flightlessness in some South African desert dwelling scarab beetles is a morphological correlate with water conservation capablities. This may also be true and part of the adaptive strategies of such flightless Galápagos beetles such as tenebrionids, carabids, and weevils.

An additional complicating factor in understanding wing reduction or loss is that it may be a sexually dimorphic phenonomenon, with females more often having greater wing reduction. This is the situation in several Galápagos *Gryllus* crickets, *Chorisoneura* cockroaches and the *Galapagia* mantid. Thayer (1992) reviews wing dimorphism in insects and suggests that there may be reproductive advantages for short wings in females in both island and continental habitats. The topic is not yet fully explored.

Roff (1994) shows that brachyptery (flightlessness) can theoretically spread rather rapidly in insect populations. While population size may become smaller with flightlessness, it allows invasion into spaces not available to macropterous forms. The genetic maintenance of population macroptery is determined by: habitat patch persistence time, energetic costs of macroptery itself, proportion of migrants from the population and probability of migrants locating other suitable habitat patches (Roff 1994).

Speciation and flightlessness. Flightless terrestrial arthropods would certainly appear to have less dispersal potential than winged ones. But primitively flightless insects (such as Collembola and silverfish) seem not to have undergone much Galápagos speciation. Rather, species proliferation has occurred in the free-living Galápagos insects that are secondarily wingless. Twenty genera (Table 4.5) of pterygote insects probably colonized in a flightless condition, but only 8 of these have undergone island multiplication to three or more species. These groups have produced an average of 2.4 species per colonization event. Another 19 genera appear to have become flightless after colonization (Table 4.6), and these show even more species proliferation, with a mean of 4.4 species per colonization event. Groups that are actively in the process of losing flight ability, such as *Chorisoneura* cockroaches, *Gryllus* crickets, and *Ataenius* and *Neoryctes* scarabs show discrete polymorphic stages in reduction of fore and (or) hind wings (Peck and Roth 1992; Cook et al. 1995; Otte and Peck 1997). So, loss of flight ability in Galápagos insects is a significant evolutionary theme. It has not always sponsored a major burst of species multiplication, but this has happened more often in groups that lost their flight ability on the Galápagos than in groups that arrived already in a flightless condition.

Speciation and island area. Is there a minimal area for an island to support the origin of new insect species? Table 5.1 shows that islands less than 1 km^2 in area do not have single-island endemic species. The pivot point for the possession of single-island endemic insects seems to be at about 1 km^2. Both Darwin

Table 4.5. List of native free living pterygote insects of the Galápagos which probably colonized in a *flightless* condition, and the estimated number of colonizations, and amount of speciation after colonization. These data give a mean of 2.4 species descending from each colonizing ancestor. The ± symbol indicates where it is not completely certain that the colonizing ancestor was flightless.

	Number of ancestral colonizations	Number of descendant species
Dermaptera		
Anophthalmolabis	1	2
Orthoptera		
Cycloptilum (Gryllidae)	3?	3?
Hygronemobius (Gryllidae)	1	1
Liparoscelis cooksonii (Tettigoniidae)	1	1
Pteronemobius (Gryllidae)	1 ?	2
Hemiptera		
Corythaica (Tingidae) (±winged species)	2	3
Darwinivelia (Mesoveliidae)	1	1
Ghinallelia (Reduviidae)	1	8
Halobates (Marine) (Gerridae)	3	3
Limnogonus (Gerridae)	1	1
Microvelia (Veliidae)	2	2
Phatnoma (Tingidae) (±winged species)	1	2
Coleoptera		
Ammophorus (Tenebrionidae)	2 ?	12
Gerstaeckeria (Curculionidae)	1 ?	5 ?
Menimopsis (Tenebrionidae)	1	1
Mystroceridius (Carabidae)	1 ?	2
Pitnus (Ptinidae)	2	3
Pseudopentarthrum (Curculionidae)	2 ?	8 ?
Scarites (Carabidae)	2	2
Stomion (Tenebrionidae)	1	11
Totals	30	73

and Wolf islands are slightly larger than this in area, and each has at least 4 insect single-island endemics. However, these are also the most isolated islands in the archipelago. It may not be easy to separate the effects of area from isolation in the speciation process.

A taxon cycle? Wilson (1961) proposed a sequence of stages in the evolution of an island fauna and called it the "Taxon Cycle." Briefly, stage I is colonization of the island; stage II represents within-island speciation and specializa-

Table 4.6. List of native free living pterygote insects of the Galápagos which probably colonized in a winged condition, and for which *secondary flightlessness* probably *evolved* or is *evolving* on the Galápagos, and estimate of number of colonizations, and amount of speciation after colonization. These data give a mean of 4.4 species descending from each colonizing ancestor.

	Number of colonizations	Number of species
Orthoptera		
Closteridea (Acrididae)	1	1
Halmenus (Acrididae)	1	4
Gryllus (Gryllidae)	2 ?	8
Mantodea		
Galapageia (females, flightless)	1	1
Blattodea		
Choristoneura (Blattellidae)	1	2
Ischnoptera (Blattellidae)	1	3
Hemiptera		
Nabis (Nabidae) (±winged ancestors?)	1	2
Phatnoma (Tingidae)	2	2
Homoptera		
Nesosydne (Delphacidae)	2 ?	6
Oliarus (Cixiidae)	1 ?	5
Philatis (Issidae)	1	20
Coleoptera		
Blapstinus (Tenebrionidae)	2	19 ?
Bythinoplectus (Pselaphidae)	1	2
Calosoma (Carabidae)	1	3
Discolus (Carabidae)	1	3 ?
Galapaganus (Curculionidae)	2 (3 ?)	10
Neoryctes (Scarabaeidae)	1	4
Pinostygus (Staphylinidae)	1	1
Pterostichus (Carabidae)	1 ?	12 ?
Scydmaenus (Scydmaenidae)	1	1
Totals	25	109

tion; stage III represents a new wave of outward colonization to other islands in the archipelago; and stage IV is outward colonization to other archipelagos. Galápagos insects are in phase I, II and III, and probably are unlikely to enter a phase IV on their own. They will naturally remain restricted to the Galápagos. Only one case of stage IV movement beyond the Galápagos is known. It is the flightless Darwin's darkling beetle *Ammophorus insularis*. This was carried to the Hawaiian Islands, probably by whaling ships, and has spread throughout all the major Hawaiian Islands, probably mostly by transport of building materials in World War II (Peck 1997). Only a few other Galápagos endemic arthropods are known either naturally or through human help to have colonized lands outside the archipelago and these examples are for Cocos Island of Costa Rica (Bickel and Sinclair 1997). That island also has the only species of Darwin's finch outside of the Galápagos.

Genus level endemics

Endemic genera probably represent an earlier time of colonization and a more prolonged period of isolation. Endemic genera are proportionally more frequent in the vertebrates and less frequent in the insects. Twenty-four endemic insect genera do exist (Table 4.4) in the over 1000 genera known. Among these endemics are some phylogenetic relicts, which have no close relatives and can be listed as paleo-endemics No mainland relative is known for the eyeless cave staphylinid *Pinostygus* of Isla Santa Cruz, nor for the *Neoryctes* dynastine scarabs which occur as four species on four islands (Cook et al. 1995). Another possible relict is the flightless locust *Halmenus* which may have descended from a very early colonization of a common ancestor shared with *Schistocerca* locusts.

Adaptive radiation

In adaptive radiation the evolutionary changes which occur in the diversification of a lineage must be ones that exploit new resource types with different morphological or physiological traits (Schluter 1996). This is a common phenomon on islands, but it is important to note that adaptive radiation is much more than just the simple species multiplication which follows genetic isolation on separate islands. Along with the morphological, physiological, and (or) behavioral changes accompanying speciation must also come changes in either or both niche and habitat as has happened in the famous textbook examples of Darwin's finches (Grant 1986). Other examples can include the striking adaptive radiation in *Scalesia* trees and shrubs, and perhaps *Opuntia* cactus. In contrast, the famous giant tortoises and less famous lava lizards have undergone much speciation or subspeciation, but I am not aware of evidence for adaptive radiation in these.

Are these few examples of adaptive radiation indicative of a generalization, or are they exceptions? How many of the monophyletic species swarms in the insects of the archipelago have undergone significant ecological, morphological, or behavioral differentiation that promotes life in a new niche or new habitat? In short, there seem to be very few examples within the insects. In the three genera of Darwin's darkling beetles (*Ammophorus, Stomion, Blapstinus*), there are some cases of congeneric species sympatry and there is some habitat separation between species based on preferences for different substrate types (sand versus ash), habitat distance from the seacoast, and elevation (Finston et al. 1997). Most *Ammophorus* species inhabit the arid zone, but two are restricted to the moist highlands of San Cristóbal and Santa Cruz islands. The eight species of *Gryllus* crickets are probably not examples because they likely sort themselves by male calling songs rather than by habitat preferences (Otte and Peck 1997). Thus, while the Galápagos are famous for having provided a classic example of the process and results of adaptive radiation in Darwin's finches, this is the exception rather than a common occurrence in Galápagos organisms.

In insect genera which have undergone extensive Galápagos speciation there is a general pattern that the ancestral species were lowland colonists, and the derivative species only later came to occupy the younger and smaller humid upland habitats. In these, the seemingly apparent mother species (1) may still occur in the lowlands, or (2) may actually be a more recent recolonization (as is possibly the case in *Calosoma* carabid beetles and *Gryllus* crickets (Otte and Peck 1997)), or (3) may now be extinct (as in *Neoryctes* scarabs; Cook et al. 1995). While some insects have changed their habitat associations, only one example is known of some change or shift in the trophic components of the niches of the insects. This is the biting horse fly *Tabanus subsimilus*, which ancestrally probably fed on the blood of warm-blooded continental vertebrates, and on the Galápagos now feeds on the cold-blooded sea turtles, tortoises, and iguanas. The only obvious evidence of niche shift in insect genera with multiple species is in their development of habitat separation along elevation and moisture gradients. This topic is worthy of additional detailed study.

Habitat shifts and subterranean arthropods

A diverse assemblage of many eyeless arthropods occurs in the extensive systems of caves and rock crevices in the basalt bedrock of the Galápagos (Peck 1990). Some ten species of arthropods such as geophilomorph centipeds, polydesmoid millipeds, soil dwelling earwigs, and darkling and carabid beetles are in phylogenetically eyeless genera which must have colonized the Archipelago in an already-eyeless condition. But at least another 23 species of eyeless terrestrial arthropods are in normally eyed groups that must have lost their eyes after colonizing the islands, and during the process of adapting to soil, litter or subterranean habitats (Peck 1990).

These species seem not to have changed their trophic niches in this process, but many have certainly moved into a habitat which is physically different from that of their above-ground (epigean) ancestors. This is a most exciting group of organisms because their changes for subterranean life can only have evolved after the origin and colonization of the islands they live on (Table 4.7), and we

Table 4.7. Summary of the groups of Galápagos eyeless insects.

Adventives

Lepidocampa zetecki (Diplura)

Nicoletia meinerti (Thysanura)

Parajapyx isabellae (Diplura)

Naturally colonizing in an already eyeless condition

Anophthalmolabis (Dermaptera); Isabela and Santa Cruz.

Menimopsis (Coleoptera, Tenebrionidae); Santa Cruz

Mystroceridius (Coleoptera, Carabidae); Floreana and Santa Cruz

Eyelessness evolved in the Galápagos in soil, litter, or caves

Anchonus persephone (Coleoptera, Curculionidae); Santa Cruz

Anurogryllus typhlops and species 2 (Orthoptera, Gryllidae); Isabela and Santa Cruz

Bythinoplectus caecus (Coleoptera, Pselaphidae); Isabela

Ischnoptera peckorum (Blattodea, Blattellidae); Santa Cruz

Oliarus hernandesi (Homoptera, Cixiidae); Floreana

Pinostygus galapagoensis (Coleoptera, Staphylinidae)Santa Cruz

Scydmaenus galapagoensis (Coleoptera, Scydmaenidae); Isabela

---

thus have a maximum time limit for dating these evolutionary changes. In some cases, the apparent eyed sister species of an eyeless species occurs on the same or neighbouring islands. Several spider examples are known (Gertsch and Peck 1992). An insect example is the cockroach sister species pair *Ischnoptera snodgrassii* (eyed, in forests of Isla Isabela) and *I. peckorum* (eyeless, in caves of Santa Cruz). In other cases, no epigean relatives are known in the Galápagos which could be a sister to the eyeless subterranean species. Examples are eyeless pholcid, gnaphosid and theridiid spiders (Gertsch and Peck 1992), the two species of eyeless *Anurogryllus* crickets, and the extraordinary staphylinid beetle *Pinostygus galapagoensis*. These are viewed as relicts of earlier colonizations. This last is the most dramatic example of any animal's change in morphology and shift in habitat since its ancestral colonization of the islands. This beetle, from an ancestral condition of a winged, visually hunting arboreal predator of insects in tropical continental forest canopies, has become an eyeless, wingless predator of subterrean invertebrates in the eternal dark of caves and deep rock crevices.

Modes of speciation

Darwin's finches are the classic textbook example of insular speciation through geographic (allopatric) isolation on separate islands of an archipelago. Most of the speciation in vertebrates and plants of the Galápagos seems to have been by this allopatric mode of speciation (Grant and Grant 1989). This is the prevalent mode of animal speciation worldwide, and needs no additional explanation here. But there are some arthropod examples which seem to suggest the alternative possibility of either parapatric or sympatric speciation.

The cases are ones in which sister species of Galápagos terrestrial arthropods have adjacent (parapatric) distributions on the same island. This occurs in litter dwelling *Pterostichus* beetles in southern Isabela, Santa Cruz, and Santiago Islands (Desender et al. 1992*a*, 1992*b*) and lycosid spiders on Santa Cruz (Maelfait and Baert 1986), as well as *Bulimulus* land snails (Chambers 1991; Coppois 1984). No apparent ecological or geographical barrier lies between these probable sister species. Several other species pairs on the same island have eyed-epigean and eyeless-hypogean sister species that are now separated by habitat, but not by geographic distance (Peck 1990; Peck and Finston 1993). This pattern of eyed and eyeless sister species also occurs in the forest and cave faunas of the Hawaiian and Canary archipelagos. This may be a pattern resulting from parapatric speciation processes acting across the strong ecological gradient of cave entrances, but to prove this over the alternative hypothesis of allopatric speciation followed by secondary contact of the sister species may be difficult or impossible. Nevertheless, the recurring pattern is exciting and is worthy of additional study.

In this context, it is interesting to note that elsewhere in the world *very isolated* and *single* high oceanic islands were colonized by some insect species that produced large (seemingly monophyletic) species swarms on the single island, in the absence of strong geographic subdivision on the island (Peck, unpublished). Such parapatric or micro-allopatric speciation processes need to be studied in more detail for island insects in general.

Extinction

The extinction of the epigean ancestors of the island paleo-endemic or relict subterranean insects was most likely caused by the exceptionally arid local climate that occurred during global glacial climates. Only if the lineage was able to occupy more moist subterranean habitats was it able to survive, as a relict, and to remain as evidence of the former existence of the epigean ancestor. Except for macrofossils of an *Azolla* water fern, these invertebrates are the only other evidence that glacial arid climates were responsible for past episodes of biotic extinction in the Galápagos. Of course, there are sub-fossil bone deposits and old collections that document vertebrate extinctions after the arrival of humans (Steadman 1986).

Contemporary extinction. Insect species extinction through human causes is probable, but no individual examples are known. Some of the adventive insects, such as *Wasmannia* fire ants (Lubin 1984) and *Polistes* wasps, are preying on or competing with indigenous and endemic insects. Introduction of non-native vertebrates has had a three-fold effect: (1) the nearly complete loss of insect host plants; such as *Opuntia* cactus on most of Floreana and San Cristóbal (eaten by feral goats and donkeys) and the concomitant loss of host specific *Gersteckaria* weevils and other insects; (2) extinction of host animals, such as the seven extinct species of endemic rice rats; and loss of their ectoparasitic fleas or lice, with the advent of black rats from ships; and (3) loss by predation; such as by mice or rats or pigs, feeding on *Neoryctes* scarab beetles or other large-bodied insects. But there is presently no strong or direct evidence of the actual extinction of an insect species on the Galápagos through an action ultimately caused by human activity.

Human caused habitat alteration has had a significant, but unmeasured, effect on the native insect populations. The clearing of large areas of *Scalesia* forest for agriculture and pastures has also affected their insect populations. The replacement of large areas of native vegetation by introduced grasses and weeds must also have had some impact. The importance of all of these introductions and alterations has not been measured or even estimated for the insects.

References

Bickel, C.J., and Sinclair, B.J. 1997. The Dolichopodidae (Diptera) of the Galápagos Islands, with notes on the New World Fauna. Entomol. Scand. **28**: 241–270.

Carson, H.L., Val, F.C., and Wheeler, M.R. 1983. Drosophilidae of the Galápagos Islands, with descriptions of two new species. Int. J. Entomol. **25**: 239–248.

Chambers, S.M. 1991. Biogeography of Galápagos land snails. *In* Galápagos marine invertebrates: taxonomy, biogeography and evolution in Darwin's Islands. *Edited by* M.J. James. Topics in Geobiology, vol. 8. Plenum Press, New York. pp. 307–325.

Chown, S.L., Pistorius, P., and Scholtz, C.H. 1998. Morphological correlates of flightlessness in southern African Scarabaeinae (Coleoptera: Scarabaeidae); testing a condition of the water-conservation hypothesis. Can. J. Zool. **76**: 1123–1133.

Cook, J., Howden, H.F., and Peck, S.B. 1995. The Galápagos Islands genus *Neoryctes* Arrow (Coleoptera: Scarabaeidae: Dynastinae). Can. Entomol. **127**: 177–193.

Coppois, G. 1984. Distribution of bulimulid land snails on the northern slope of Santa Cruz Island, Galápagos. Biol. J. Linn. Soc. **21**: 217–227.

Desender, K., Baert, L., and Maelfait, J.-P. 1992a. El Niño events and establishment of ground beetles in the Galápagos Archipelago. Bull. Inst. R. Sci. Nat. Belg. Entomol. **62**: 67–74.

Desender, K., Baert, L., and. Maelfait, J.-P. 1992b. Distribution and speciation of carabid beetles in the Galápagos Archipelago (Ecuador). Bull. Inst. R. Sci. Nat. Belg., Entomol. **62**: 57–65.

Eldredge, L.G., and Miller, S.E. 1995. How many species are there in Hawaii? Bishop Mus. Occas. Pap. No. 41. pp. 1–18.

Eldredge, L.G., and Miller, S. E. 1997. Numbers of Hawaiian species: supplement 2. Bishop Mus. Occas. Pap. No. 48. pp. 3–22.

Finston, T., and Peck, S.B. 1997. Genetic differentiation and speciation in *Stomion* (Coleoptera: Tenebrionidae): flightless beetles of the Galápagos Islands, Ecuador. Biol. J. Linn. Soc. **61**: 183–200.

Finston, T., Peck, S.B., and Perry, R.B. 1997. Population density and dispersal ability in Darwin's Darklings: flightless beetles of the Galápagos Islands, Ecuador. Pan-Pac. Entomol. **73**: 110–121.

Gertsch, W.J., and Peck, S.B. 1992. The pholcid spiders of the Galápagos Islands, Ecuador (Araneae: Pholcidae). Can. J. Zool. **70**: 1185–1199.

Grant, P.R. 1986. Ecology and Evolution of Darwin's finches. Princeton University Press, Princeton, NJ. 458 pp.

Grant, P.R., and Grant, B.R. 1989. Sympatric speciation and Darwin's finches. *In* Speciation and its consequences. *Edited by* D. Otte and J.A. Endler. Sinauer Associates, Sunderland, MA. pp. 433–457.

Lubin, Y.D. 1984. Changes in the indigenous fauna of the Galápagos Islands following invasion by the little red fire ant, *Wasmannia auropunctata*. Biol. J. Linn. Soc. **21**: 229–242.

Maelfait, J.-P., and Baert, L. 1986. Observations sur les Lycosidae del Iles Galápagos. Mem. Soc. R. Belg. Entomol. **33**: 139–142.

Otte, D., and Peck, S.B. 1997. New species of *Gryllus* (Orthoptera: Grylloidea: Gryllidae) from the Galápagos Islands. J. Orthoptera Res. **6**: 161–174.

Peck, S.B. 1990. Eyeless arthropods of the Galápagos Islands, Ecuador: composition and origin of the cryptozoic fauna of a young tropical oceanic archipelago. Biotropica **22**: 366–381.

Peck, S.B. 1997. *Ammophorus insularis* in Hawaii: A Galápagos Islands species immigrant to Hawaii (Coleoptera: Tenebrionidae). Bishop Mus. Occas. Pap. No. 49. pp. 26–29.

Peck, S.B., and Finston, T. 1993. Galápagos Islands troglobites: the questions of tropical troglobites, parapatric distributions with eyed-sister-species, and their origin by parapatric speciation. Mem. Biospeol. **20**: 19–37.

Peck, S.B., and Roth, L.M. 1992. Cockroaches of the Galápagos Islands, Ecuador, with descriptions of three new species (Insecta: Blattodea). Can. J. Zool. **70**: 2202–2217.

Roff, D.A. 1991. The evolution of flightlessness in insects. Ecol. Monogr. **60**: 389–421.

Roff, D.A. 1994. Habitat persistence and the evolution of wing dimorphism in insects. Am. Nat. **144**: 772–798.

Schatz, H. 1998. Oribatid mites of the Galápagos Islands — faunistics, ecology and speciation. Exp. Appl. Acarol. **22**: 373–409.

Schluter, D. 1996. The ecological causes of adaptive radiation. Am. Nat. **148**: 540–564.

Steadman, D.W. 1986. Holocene vertebrate fossils from Isla Floreana, Galápagos. Smith. Contrib. Zool. **413**: 1–103.

Thayer, M.K. 1992. Discovery of sexual wing dimorphism in Staphylinidae (Coleoptera): "*Omalium*" *flavidum* and a discussion of wing dimorphism in insects. J. N.Y. Entomol Soc. **100**: 540–573.

Wilson, E.O. 1961. The nature of the taxon cycle in the Melanesian ant fauna. Am. Nat. **95**: 169–193.

Chapter 5
Patterns: distributional and ecological determinants of composition or structure of an insect fauna

A great many physical and biological factors on continental landmasses influence the distribution of species and the faunal assemblages composed of these species. It is often difficult or impossible to determine the relative importance of individual factors. There is a long-standing debate in ecology about what controls or determines community composition. There are two major categories of explanation: (1) non-equilibrium determinants and (2) equilibrium theory. In non-equilibrium theory, the determinants of diversity are environmental processes and these determine community composition and structure through the processes of disturbance and the variables of habitat heterogeneity. The processes of recovery from the last disturbance generate diversity in the community, not equilibrium conditions. The processes set the range of possiblilites for random colonization, reproduction, and survival, and the variation that results in the return of an assemblage to a condition that may differ from the original assemblage (Riece 1994; Sousa 1984). Recent ecological attempts to encompass this view are called patch dynamics and disturbance theory. These have been studied in rain forests, coral reef fishes, stream invertebrates, chalk grasslands, and pine savannas (Reice 1994), but I am not aware that these approaches have been applied to oceanic islands. With the frequent environmental disturbances experienced in the volcanic eruptions and natural fires of the Galápagos, they should be a good model system for the application of such approaches.

The alternative, an equilibrium theory of community composition, is presently the dominant ecological paradigm. This emphasizes the role of biological interactions between species, especially in the processes of competition and predation. These processes are principal in determining what subset of species actually exists in an area, from the larger set of species that are possible. The implication is that overall species composition and abundances are stable through time, and will return to the original structure after disturbance. This view has been preeminant in modern attempts to analyze and understand the composition of biotic assemblages on islands.

On islands, it is often easiest to measure or quantify physical environmental factors and their relationship to the distributional patterns of the biota. A synthesis of the relative importance of some of these was presented as a general theory of equilibrium island biogeography by MacArthur and Wilson (1967). Island colonization and the ultimate biotic diversity on an island was shown to be controlled or at least partly influenced by island area and isolation, and by biotic immigration and extinction rates. These relatively easily measured factors are viewed as proxies for measures of more obscure biological interactions. A vast literature has analyzed many groups of organisms for these and other influences on island species numbers (Borges and Brown 1999), but few studies have addressed the factors which may influence the diversity of a large suite of Pacific island insects. For some Costa Rican offshore islands, Janzen (1983: 637) observed a depression of insect density and diversity caused by a variety of possible, but unmeasured, reasons. The most extensive study has analyzed the physical geographical factors which are correlated with the rich native insect fauna of the Hawaiian Islands (Peck et al. 1999).

Physical Determinants of Insect Diversity

The effect of age, present and past area, elevation, isolation, habitat diversity, and so on in determining species diversity of plants and birds on the Galápagos have been investigated in various studies (Connor and Simberloff 1978; Hamilton and Rubinoff 1963, 1964, 1967; Hamilton et al. 1963, Johnson and Raven 1973; Simpson 1974; Thornton 1967). Unfortunately, it is still premature to assess the effects of these factors on the entire diversity of all the terrestrial arthropods or of even all the insects. But at least the rich species diversity of the insects may ultimately provide a better analytical data base for such questions than have the equivocal results from the lower diversities of the birds and vascular plants. I have sumarized the data which appears in the following chapters on the small insect orders to investigate some of the more obvious possible determinants of the insect assemblages. A summary of what follows is that in the Galápagos an island's insect species diversity is positively correlated with its area, elevation, and ecological complexity (data from Table 1.2).

Island area and insect diversity. It is generally found that larger islands have more species and genera. This is not because of their larger area *per se*, but that greater area increases the chances for the occurrence of more available habitats and microhabitats. This is certainly the case in the insects of the Hawaiian Islands (Peck et al. 1999). Both specific and generic diversity of native Galápagos insects is clearly positively correlated with island area (Fig. 5.1, data from Table 5.1), as is predicted by theory. There is also a significant positive relationship between an island's area and the number of endemic species limited to it (Fig. 5.2). There is a non-significant increase in numbers of species in a genus with increasing island area (Fig. 5.3). These all show that larger islands have more species and genera. But increased area does not clearly promote an increase in species numbers within a genus on a single island in the Galápagos. Differing species compositions between islands can lead to differing island ecosystem properties (Wardle et al. 1997), but this has not been studied in the Galápagos.

Figs. 5.1–5.4. Fig. 5.1. Simple linear regression of the total number of native smaller order insect species against island area. This and the following regressions use data only for the better sampled principal islands as given in tables 1.1 and 1.2. This shows that larger islands have more species. Fig. 5.2. Simple linear regression of the number of single-island endemic insect species (smaller orders only) against island area for the Galápagos Islands. This shows that larger islands have more endemic species. Fig. 5.3. Simple linear regression of the species per genus ratio of native smaller order insects against island area. Fig. 5.4. Simple linear regression of maximum island elevation against total number of native species for the smaller order insects. This shows that higher islands have more species.

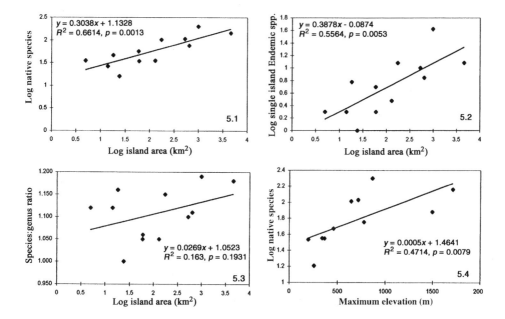

Island elevation and insect diversity. There is a strong correlation of island elevation and insect species diversity in Hawaii (Peck et al. 1999). Island elevation can influence species diversity because increased elevation should allow more habitat diversity through the existence of a greater variety of climatic conditions. In the Galápagos, temperature declines and rainfall increases with increased elevation. As predicted, there is a significant increase in insect species number with increasing island elevation (Fig. 5.4). This also occurs and is statistically significant in single-island endemics (Fig. 5.5), in spite of each island's smaller area at higher elevations, and reduced time for formation of new species in the higher elevations. There is not a significant increase in species per genus with increased elevation (Fig. 5.6). In contrast, Schatz (1998) found that Oribatida mite species richness is highest at the higher humid elevations and not the lower arid elevations.

Island age and insect diversity. It is to be expected that older islands have had more time to assemble and evolve species. If other factors were equal, older

Table 5.1. Native generic, specific, and single-island endemic insect diversity in the smaller insect orders, arranged by increasing island size.

Island	Total Genera	Total Species	Single-island Endemics	Species: Genus Ratio
Caamaño	1	1	0	1.0
Beagle	2	2	0	1.0
Campeón	4	4	0	1.0
Plazas Sur	10	10	0	1.0
Eden	4	4	0	1.0
Daphne Major	5	5	0	1.0
Gardner at Floreana	11	12	0	1.09
Darwin	11	11	4	1.0
Bartolomé	11	11	0	1.0
Tortuga	5	5	0	1.0
Wolf	19	19	4	1.0
Seymour	23	23	1	1.0
Rábida	33	37	2	1.12
Genovesa	25	28	2	1.12
Pinzón	44	52	6	1.18
Santa Fé	17	17	1	1.0
Baltra	24	24	1	1.0
Pinta	58	61	5	1.05
Española	35	36	2	1.03
Marchena	37	39	3	1.05
Floreana	99	114	12	1.15
San Cristóbal	99	116	5	1.17
Santiago	96	106	5	1.10
Fernandina	74	82	7	1.11
Santa Cruz	201	240	42	1.19
Isabela	131	155	12	1.18

islands would be expected to have a higher species diversity. This has been found for native arthropods in pastures in the Azores Islands (Borges and Brown 1999). While the areas of the islands of the Galápagos are accurately known, there is often a substantial difference between estimated minimum and maximum ages of each island (Table 1.2). No significant relationship is found between number of species, or species:genus ratios for either maximum or minimum ages (Figs. 5.7–5.9). There is actually a loss of species on old islands, probably caused by the loss of area and habitats and extinction of species as the island subsides and erodes in its old age.

Figs. 5.5–5.8. Fig. 5.5. Simple linear regression of maximum island elevation against number of single-island endemic species for the smaller order insects of the Galápagos Islands. Fig. 5.6. Simple linear regression of maximum island elevation against species:genus ratio for the smaller order insects of the Galápagos archipelago. Fig. 5.7. Simple linear regression of maximum estimated island age against total number of native smaller order insect species of the Galápagos Islands. This and the following figures show that older islands do not have more, but actually fewer species. Fig. 5.8. Simple linear regression of maximum estimated island age against total number of single island endemic smaller order insect species of the Galápagos Islands .

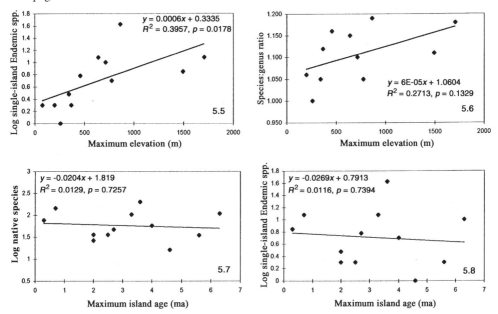

Vegetational zonation and diversity. Terrestrial communities in the Galápagos are usually characterized according to the elevation-related (precipitation and temperature controlled) zonation of the flora (Wiggins and Porter 1971). The archipelago may possess the strongest or most compressed floristic zonation to be found anywhere in the world, passing through its six major vegetation zones (Fig. 1.5) in an elevational rise of only about 700 m; the littoral, arid, transition, humid forest, evergreen shrub, and above-treeline fern-sedge ("pampa") zones. The best study to date on the zonation of terrestrial invertebrates is that of Coppois (1984) on the remarkably diverse bulimulid land snails on Santa Cruz. Spider communities have also been characterized by vegetation zones (Baert et al. 1991). Beetle diversity also seems to show some zonation, with fewer species being known from the higher elevations (Peck and Kukalova-Peck 1990).

The greatest insect species diversity is in the arid zone, which also has the largest habitat area in the archipelago (Tables 5.2 and 5.4). It is unfortunate that

Figs. 5.9–5.12. Fig. 5.9. Simple linear regression of species:genus ratio against maximum estimated island age for the smaller order insects of the Galápagos Islands. A similar regression using minimum island age as the dependent variable gives results of comparable significance. Fig. 5.10. Simple linear regression of island ecological complexity (as measured by number of major life zones) against number of native species (excluding Islas Darwin and Wolf). Fig. 5.11. Simple linear regression of island ecological complexity (as measured by number of major life zones) against number of single island endemic species. Fig. 5.12. Simple linear regression of island ecological complexity (as measured by number of major life zones) against species:genus ratio.

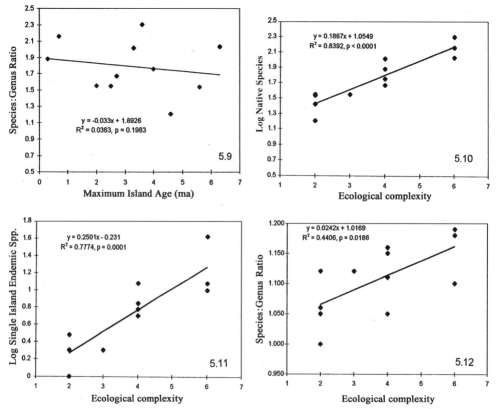

figures are not available for the total areas of the different vegetation zones and the areas of barren lava on each island. It would be very informative to include these in a future analysis. A transect study of Volcan Alcedo on Isabela also found the insect species diversity to be highest at lower elevations (Desender et al. 1999).

It might normally be expected that the terrestrial arthropods would have greater diversity and abundance in the more favorable habitats of the humid zone of the higher islands (Jackson 1985; p. 203) even though these habitats are markedly smaller and younger. This has been found to be true for Oribatida mites (Schatz 1998). But this does not occur in the insects. Their diversity decreases with elevation as the area of the ecological zone decreases, even though more favorable conditions (such as more moisture) occur in the humid forest zone (Table 5.2). The data for the transition zone seems to show that it has a lower insect diversity than the humid zone, but I suspect that this is probably an artifact of inadequate sampling.

Trophic categories. Becker (1975) has noted that island colonization by beetles worldwide is more frequent in predator species,which are more likely to be trophic generalists, than by herbivores, which are more likely to be trophic specialists. This is the pattern in Galápagos beetles (Peck and Kukalova-Peck 1990). But Becker (1992) did not find this pattern in true bugs worldwide, and this is also the Galápagos pattern. The reasons for the differences are not clear. Galápagos data for all the small orders show that there are generally more species in each life zone (Table 5.3) of all subgroups of herbivores than of predators and scavengers combined and when trophic category is also compared to the species grouped as endemic, indigenous, and adventive (Table 5.4).

Ecological richness and species diversity. Vegetational zones do have some predictive power for understanding the insect fauna. The greatest diversity is in the arid lowlands, the largest ecological zone of the islands (Table 5.2, 5.3). The insect fauna is generally (but not strongly) zonal. Most insect species occur in more than one vegetational zone, as do most plant species. The strongest insect zonation is for halophilic species in the littoral zone, and hygrophilous species in the humid forest-pampa zones. The lack of strong insect zonation is because most species are winged and can follow favorable climatic conditions up or down a mountain. Flightless species are less able to do this. An analysis of prevalence of flightlessness with zonation shows that there is a pattern of flightless xerophilous species in the lowlands and flightless hygrophilous species in the seasonally moist highlands.

Host-specific plant-feeding insects could be expected to have the zonation of their hosts, but almost all Galápagos phytophages seem to feed on more than one species of host plant. Host specificity, to be expected in groups which elsewhere are usually monophagous or stenophagous plant-feeders, such as mirid and lygaeid bugs, is lacking in Galápagos phytophages and this was first noted by Usinger (1972: 278). Our studies have found no additional evidence for host specificity in native phytophagous insects other than in *Gerstaeckeria* weevils on *Opuntia* cactus and Monarch butterflies on *Asclepias* milkweed.

Is there a relationship between the ecological diversity of an island and its insect species diversity? This would be expected, especially in larger and higher islands, because these are correlated with diversity. A very strong relationship was found for the orthopteroid insects (Peck 1996). Schatz (1998) found that

Table 5.2. Numbers of all small order insect species by distributional categories and their occurrence in the different vegetation life zones of the Galápagos (note that most species occur in more than one life zone).

					Ecological Life Zone				
Distribution Category	Littoral	Arid	Transition	Humid Forest	Evergreen Shrub	Pampa	Inversion	Urban	Agriculture
Single-island Endemic	10	49	15	35	8	14	0	0	13
Archipelago Endemic	25	80	56	62	40	41	3	1	17
Indigenous	50	89	64	77	45	48	3	10	34
Adventive	22	69	45	41	20	20	2	14	31
Totals	107	287	180	215	113	123	8	25	95

Table 5.3. Numbers of all small order insect species by trophic categories and their occurrence in the different vegetation life zones of the Galápagos Islands.

	Littoral	Arid	Transition	Humid Forest	Evergreen Shrub	Pampa	Inversion	Urban	Agricultural
Predators	31	52	36	37	27	28	1	5	15
Scavengers	28	51	48	64	38	43	1	13	45
Fungivores	4	10	5	3	1	1	0	0	0
Herbivores (Sap)	30	118	52	72	26	28	4	3	23
Herbivores (Leaves)	7	38	28	28	14	15	0	0	8
Herbivores (Wood)	2	3	2	2	0	1	0	1	1
Herbivores (Seeds)	4	14	8	8	7	7	2	1	3
Endoparasites	1	0	0	0	1	0	0	2	0
Ectoparasites	0	0	1	1	0	0	0	0	0
Totals	107	286	180	215	114	123	8	25	95

Table 5.4. Distribution and trophic categories of small order insect species of the Galápagos
Islands.

Category	Single-island Endemics	Archipelago Endemics	Indigenous	Adventive	Total
Predators	10	17	40	9	76
Scavengers	21	24	38	35	118
Fungivores	0	0	9	2	11
Herbivores (Sap)	69	55	25	63	212
Herbivores (Leaves)	12	10	24	7	53
Herbivores (Wood)	0	1	3	0	4
Herbivores (Seeds)	2	8	3	2	15
Endoparasites	0	1	1	0	2
Ectoparasites	0	1	0	4	5
Totals	114	117	143	122	

mite species richness on an island depends on the number of available habitats, and not island area. When all of the smaller insect orders are combined, and ecological diversity of an island is measured as a sum of its major ecological zones, there is a significant relationship with species numbers, single island endemics and the species:genus ratios (Figs. 5.10–5.12).

Multiple regression analyses. When interpreting the biological significance of the results of the previous simple linear regression analyses (Figs. 5.1–5.9), one must acknowledge that two island parameters, maximum island elevation and island area, are strongly correlated. This pattern is undoubtedly related to the fact that the Galápagos are volcanic islands, and volcanoes grow "up" and spread "out" in unison. In addition, ecological complexity is strongly related to both of these variables (Figs. 5.13–5.15). While all three of these island variables were statistically significant predictors of insect diversity patterns, their correlation makes it difficult to conclude whether both or either of the variables actually support the tested hypotheses of equilibrium island biogeography. Another problem with relying solely on the above simple linear regression analyses relates to the possibility of the masking of variables. This occurs when one predictor variable exhibits such a strong influence on the dependent variable that less strongly associated predictive relationships are not detectable; relationships are essentially lost in the noise created by other influences.

Because of these two problems with simple linear regression analyses, the same data were analyzed using a multiple linear regression (MLR) approach. Multiple linear regression proceeds by first accounting for the variability due to the strongest predictors and then allows for detection of more subtle relationships. This avoids the problems of correlation and masking. MLR decreases the

Figs. 5.13–5.15. Multiple linear regression examinations of correlated determinants of insect species diversity in the Galápagos. Area, elevation, and ecological complexity are not independent from each other. These generally yield similar results to the simple linear regressions.

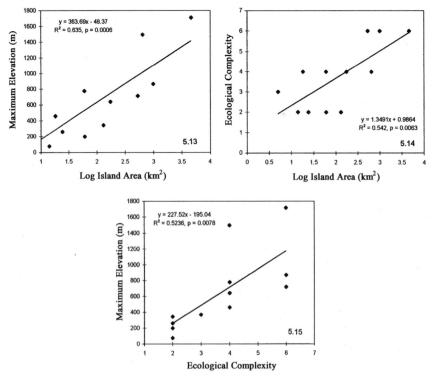

probability of concluding by chance that a relationship is evident when it is not. The results of the MLR analysis (Figs. 5.13–5.15, Table 5.5) are somewhat different than those coming from the single linear regressions, probably relating to both the correlation and masking of predictors. Ecological complexity was found to be a significant predictor of all three measures of insect diversity (also observed in the simple regression approach). However, while island area and maximum elevation were both significant positive predictors of the number of single island endemic species and of the number of indigenous species, neither explained variation beyond that of ecological complexity in the MLR approach. While no variables were found to be significant predictors of the species:genus ratio in the simple linear regression models, the MLR analysis yielded a three-term model: ecological complexity (positive), island age (negative), and island elevation (negative). The high level of correlation between the variables of ecological complexity, island age, and island elevation undoubtedly accounts for the discrepancy between the two regression approaches. Though one could argue that the MLR analysis provides the most reliable support for hypotheses, each analysis has its own unique merits for assisting the biological interpreta-

Table 5.5. Multiple linear regression analyses testing the relationships between small order insect fauna (number of indigenous species, number of single-island endemic species, and species:genus ratios) and four island parameters (maximum island elevation, maximum island age estimate, ecological complexity, and island area). Models were built using the forward-stepwise procedure with addition of terms based on alpha = 0.05.

A. Regression Summary for Dependent Variable: Log Indigenous Species

$R = 0.9161$ $R^2 = 0.8392$ Adjusted $R^2 = 0.8231$

$F(1,10) = 52.196$ $p < 0.0001$ Std. Error of estimate: 0.1374

	Coefficient	St. Error	$t(10df)$	p-level
Intercept	1.0549	0.1047	10.0724	<0.0001
Ecological Complexity	0.1867	0.0258	7.2247	<0.0001

B. Regression Summary for Dependent Variable: Log Single-Island Endemics

$R = 0.8817$ $R^2 = 0.7774$ Adjusted $R^2 = 0.7552$

$F(1,10) = 34.929$ $p < 0.0002$ Std. Error of estimate: 0.2250

	Coefficient	St. Error	$t(10df)$	p-level
Intercept	–0.2310	0.1715	–1.3473	0.2076
Ecological Complexity	0.2501	0.0423	5.9101	0.0001

C. Regression Summary for Dependent Variable: Species:genus Ratio

$R = 0.8853$ $R^2 = 0.7838$ Adjusted $R^2 = 0.7028$

$F(3,8) = 9.6693$ $p < 0.0049$ Std. Error of estimate: 0.0318

	Coefficient	St. Error	$t(8df)$	p-level
Intercept	1.0809	0.0308	35.0680	<0.0001
Ecological Complexity	0.0473	0.0106	4.428	0.0022
Maximum Island Age	–0.0265	0.0074	–3.5593	0.0074
Maximum Elevation	–0.0001	0.0001	–2.6420	0.0296

tion of underlying relationships. To assume that the MLR analysis necessarily provides more accurate insight would assume that accuracy can be based merely on a slightly stronger correlation of one predictor variable compared to another.

Vegetational zone changes through time. It must be realized that the vegetational zones have not been stable through time, but have slowly moved up or down in elevation in response to climatic conditions. Colinvaux and Shofield (1976a, b), using pollen and spores recovered from cores of lake sediments and bogs, have shown that the present woody flora of the highland humid zones was markedly reduced or almost absent about 10,000 yr BP, at the end of the Pleistocene. This means that during Pleistocene glacials the arid-zone was much larger (especially with lower sea levels (Fig. 1.2)) and the forested moist-zone very much smaller than at present. The present moist-zone forest plant community may have been mostly assembled in only the past 10,000 years and some or many of the hygrophilous plant species may be recent colonists (Johnson and

Table 5.6. The number of endemic small order insect species and genera with endemic species occurring in each vegetation life zone in the Galápagos Islands. Also given are the species:genus ratios, mean zonal breadth of the insects that occur within each life zone, and percent zonal endemicity (percentage of species in each life zone that are restricted to that life zone). The comparatively high figures for the agriculutral zone show that much of the native fauna can still exist and be found in this disturbed area.

Life Zone	Total Endemic Species	Archipelago Endemics	Single-island Endemics	Genera with Endemic Species	Species: Genus Ratio	Mean Zonal breadth	Percent Zonal Endemicity
Littoral	34	24	10	27	1.25	3.47	23.53
Arid	129	80	49	70	1.84	2.77	44.96
Transition	71	56	15	52	1.36	4.17	7.04
Humid forest	97	62	35	64	1.52	3.42	27.84
Evergreen shrub	48	40	8	38	1.26	4.94	2.08
Pampa	55	41	14	43	1.28	4.27	14.55
Inversion	3	3	0	3	1.00	5.67	0.00
Urban	1	1	0	1	1.00	8.00	0.00
Agricultural	30	17	13	28	1.07	3.93	23.33

Raven 1973). The question of such a post-Pleistocene faunal assembly has yet to be addressed to arthropods, but it can be predicted that moist-zone habitats should have a lower number of endemic species, and a higher number of indigenous (recently colonizing) species per unit area of habitat. A preliminary analysis suggests that this is true for the beetles (Peck and Kukalova-Peck 1990) and for the smaller insect orders because their mean body size is smaller in the humid forest than in the arid zone. This suggests the Recent accumulation of indigenous species in the humid zone from South or Central American source areas by aerial transport. An analysis of indigenous and endemic species in relationship to vegetational zones shows a pattern of greatest diversity in the arid zone, with the humid zone having the second greatest amount of diversity (Tables 5.2 and 5.6). The lower numbers for the transition zone may be a collecting artifact, and this zone should be expected to be the second most diverse in insect species.

Stochastic processes in colonization and distribution. Nature is seldomly strictly predictable or linear in the dynamics of the life cycle of a group of islands, even if the islands themselves are relatively linear in age or geography. As an example, one would predict that insect colonization was first to San Cristóbal and Española, which are the oldest and most easterly islands, and that the other islands were colonized sequentially northwestward through time. Exceptions to these predicted patterns do exist. Some plant and bird species are missing from islands located between two other islands on which they do occur. This seem-

Table 5.7. Number of all small order insect species collected during each month of the year in the Galápagos Islands arranged by trophic categories (all available data combined). This shows that most species have been sampled (are present) from January–June, which are the rainy months, and when most insect collecting activity has occurred.

	Jan.	Feb.	Mar.	Apr.	May	June	July	Aug.	Sept.	Oct.	Nov.	Dec.
Predators	27	30	57	55	50	31	16	12	8	9	5	13
Scavengers	39	50	55	52	56	26	11	8	8	7	7	8
Fungivores	2	4	8	6	7	2	0	0	0	0	0	0
Herbivores (Sap)	43	75	50	59	60	41	17	4	3	2	4	29
Herbivores (Leaves)	9	21	36	32	35	22	7	4	6	3	3	2
Herbivores (Wood)	1	2	1	1	3	2	0	0	0	0	0	1
Herbivores (Seeds)	3	4	8	11	9	9	7	3	5	2	2	2
Endoparasites	1	1	1	0	0	0	0	0	0	0	0	0
Ectoparasites	1	3	2	0	0	0	0	0	0	0	0	0
Totals	126	190	218	216	220	133	58	31	30	23	21	55

ingly random or unexpected distribution pattern also occurs in some of the fairly well known groups of insects. It shows the lack of absolute predictability in present distributions through the randomness of the processes of either past dispersal, or colonization success, or extinction. But still, neighbor islands are more likely to share endemic species. This is clear in a number of shared insect species limited to island pairs such as Darwin and Wolf, and Marchena and Pinta. The isolation of Genovesa is evident in its failure to be colonized by flightless *Ammophorus* beetles and other insect groups. Pinzón is famous for not having the widespread palo santo tree (*Bursera graveolens*) (Thornton 1971), but this island's insects are not well enough known to evaluate a pattern of insect absence there.

Environmental conditions regulate periods of insect activity. With the arrival of the Galápagos rainy season insect activity increases and there are large and noteworthy insect outbreaks, which seem to be short-lived plagues, of *Calosoma* ground beetles, *Camponotus* ants, *Disclisioprocta stellata* (a geometrid moth), various sphinx moths, and other insects. These are best noticed at lights at night. These are mass emergences that are environmentally triggered, but they also occur in coastal mainland Ecuador and seasonal forests elsewhere in Central and South America, so they are not a unique island feature. The seasonal flush of high caterpillar abundance, changing from one of near absence, is similar to that in the seasonal deciduous forest of Costa Rica (Janzen 1988). Table 5.7 shows that most adult insect species are present or active during the rainy months of

January–June. Seasonal patterns should be documented by a program of regular quantitative sampling, because the present data are an artifact of the activity of collectors. There are no Galápagos studies that have followed the abundance of an insect through an entire annual cycle, or even a seasonal cycle.

El Niño rains cause a pulse in plant production and a change in terrestrial arthropod populations on arid islands in the Gulf of California (Polis et al. 1997, 1998). Insect abundance tracked the increase in plant productivity and crashed in the following normal dry year. Insect species composition was changed from the normal dominance of detritivores and scavengers to a prevalence of herbivores during the El Niño to a return to dominance of detritivores and scavengers. The slow release of plant and detrital biomass reserves were thought to have an effect for years after the El Niño had passed. All these events and changes can be expected in Galápagos insect assemblages but have not been studied.

Community or Guild Structure. Assemblages of organisms are grouped into interacting communities by a sharing of similar requirements for survival and reproduction. Because plants cannot move their locations in the life span of an individual, plant assemblages are most useful for the definition of communities. Insects usually have loosely structured associations with their plant communities. If this structure in the Galápagos is deterministic (evolved) or is instead a random assemblage can only be answered by comparison with mainland communities.

Are Galápagos insect assemblages differently structured? There is an absence of studies on the South American mainland for comparison with the Galápagos. There are certainly fewer interacting species of plants, herbivores, predators, parasitoids and scavengers. Many complexly structured Neotropical guilds are absent, such as that of Scarabaeine dung beetles (Peck and Forsyth 1982).

There are few close (species specific) plant-insect relationships. An unusually large proportion of flowering plants do not require insect pollination (McMullen 1993). Conspicuous flowers are notable by their absence, and the only pollinating bee is *Xylocopa darwini*, a carpenter bee. Flies, beetles, butterflies, and moths do most of the additional insect pollinating. A very large proportion of the flora is self-pollinating (McMullen 1987). My observations suggest that most plant feeding insects are generalists in their selection of host plants. I have found no evidence of a radiation of phytophagous insect species on the endemic plant genera such as *Scalesia* or *Darwiniothamnus*, or unique to pure stands of zonal endemics such as *Scalesia pedunculata* trees or *Miconia robinsoniana* shrubs. Arid zone dominants, such as palo santo trees (*Bersera*) or *Opuntia* cactus, could also be keystone species for an insect assemblage, but this too is unstudied.

Trophic constraints have been important in limiting the establishment of colonists such as bruchid seed weevils and the Monarch butterfly, but are not yet evident as limitations of other aspects of community assembly. Compton et al. (1994) have shown that the absence of some fig wasps on Anak Krakatau island limits the fruit set on some fig tree species, and can thus affect populations of frugivorous birds or bats. Such insect-based food chain constraints are not yet recognized in the Galápagos, except in adventive pest insects. The little red fire ant (*Wasmannia auropunctata*) seriously reduces the diversity of arachnids and insects where it occurs (Lubin 1984). The adventive vespid wasps *Polistes versicolour* and *Brachygaster lechuagana* prey voraciously on lepidopteran larvae to feed their own larvae, and may thus be serious competitors for these major food items with many of the passerine birds, including Darwin's finches.

If community equilibrium or species saturation exists it would be most evident in the older islands. But the uplands of the oldest islands of San Cristóbal and Floreana have been severely disturbed by human activity, and some insect species extinction may have occurred, but is unrecognized. Equilibrium species numbers in arid zone communities could be studied through the age sequence of islands from the oldest at 3 myr of age right up to those of the youngest Recent lava flows.

Primary succession occurs as new lava flows and ash fields weather and are colonized by plants and animals. This is a dominant ecological theme in volcanic landscapes worldwide. For instance, the Sierra Negra volcano on Isla Isabela has undergone a 90% resurfacing by new lava flows in the past 4500 years (Reynolds et al. 1995), and probably many subsequent recolonizations by its plants and animals. There are only vague notions of this succession process with the Galápagos arthropod fauna. Secondary succession can start with the destruction of a community by fire associated with a volcanic eruption, or by human-caused fires. Abedrabbo (1988) studied invertebrate recolonization in the moist pampa zone (above 750 m) of Volcan Sierra Negra, Isla Isabela, following the 20,000 ha fire of 1985. The process of secondary succession seemingly parallels primary succession, but occurs in a shorter time span. With the return of fern-grass-sedge vegetation on Sierra Negra came the pioneer colonists such as predatory spiders and rove and carabid beetles, omnivores such as ants, and herbivores such as crickets and grasshoppers. Such recovery is much slower at lower elevations, where there is less rainfall. Continued volcanic activity, especially on Marchena, Isabela, and Fernandina will continue to provide new lava and ash substrate for primary succession.

References

Abedrabbo, S. 1988. Efectos del incendio de 1985 sobre los invertebrados en Sierra Negra, Isla Isabela, Galápagos. Tesis de licenciatura. Pontifical Catholic University of Ecuador, Quito. 232 pp.

Baert, L., K. Desender, and Maelfait, J.-P. 1991. Spider communities of Isla Santa Cruz (Galápagos Eucador). J. Biogeogr. **18**: 333–340.

Becker, P. 1975. Island colonization by carnivorous and herbivorous Coleoptera. J. Anim. Ecol. **44**: 893–906.

Becker, P. 1992. Colonization of islands by carnivorous and herbivorous Heteroptera and Coleoptera: effects of island area, plant species richness, and 'extinction rates'. J. Biogeogr. **19**: 163–171.

Borges, P.A.V., and Brown, V. K. 1999. Effect of island geological age on the arthropod species richness of Azorean pastures. Biol. J. Linn. Soc. **66**: 373–410.

Colinvaux, P.A., and Schofield, E.K. 1976*a*. Historical ecology in the Galápagos Islands. I. A Holocene pollen record from El Junco Lake, Isla San Cristóbal. J. Ecol. **64**: 989–1012.

Colinvaux, P.A., and Schofield, E.K. 1976*b*. Historical ecology of the Galápagos Islands. II. A Holocene spore record from El Junco Lake, Isla San Cristóbal. J. Ecol. **64**: 1013–1026.

Compton, S.G., Ross, S.J., and Thornton, I.W.B. 1994. Pollinator limitation of fig tree reproduction on the island of Anak Krakatau (Indonesia). Biotropica **26**: 180–186.

Connor, E.F., and Simberloff, D. 1978. Species number and compositional similarity of the Galápagos flora and avifauna. Ecol. Monogr. **28**: 219–248.

Coppois, G. 1984. Distribution of bulimulid land snails on the northern slope of Santa Cruz Island, Galápagos. Biol. J. Linn. Soc. **21**: 217–227.

Desender, K., Baert, L., Maelfait, J.-P., and Verdyck, P. 1999. Conservation on Volcán Alcedo (Galápagos): terrestrial invertebrates and the impact of introduced feral goats. Biol. Conserv. **87**: 303–310.

Hamilton, T.H., and Rubinoff, I. 1963. Isolation, endemism, and multiplication of species in Darwin Finches. Evolution **17**: 388–403.

Hamilton, T.H., and Rubinoff, I. 1964. On models predicting abundance of species and endemics for the Darwin Finches in the Galápagos Archipelago. Evolution **18**: 339–342.

Hamilton, T.H., and Rubinoff, I. 1967. On predicting insular variation in endemism and sympatry for the Darwin Finches in the Galápagos Archipelago. Am. Nat. **101**: 161–171.

Hamilton, T.H., Rubinoff, I., Barth Jr, R.H., and Bush, G.L. 1963. Species abundance: natural regulation of insular variation. Science **142**: 1575–1577.

Jackson, M. 1985. Galápagos: A Natural History Guide. University Calgary Press: Calgary, Canada.

Janzen, D.H. 1983. Insects. *In* Costa Rican Natural History. *Edited by* D.H. Janzen. Univ. Chicago Press, Chicago, IL. pp. 619–645.

Janzen, D.H. 1988. Ecological characterization of a Costa Rican dry forest caterpillar fauna. Biotropica **20**: 120–135.

Johnson, M.P., and Raven, P. 1973. Species number and endemism: The Galápagos Archipelago revisited. Science **179**: 893–895.

Lubin, Y.D. 1984. Changes in the indigenous fauna of the Galápagos Islands following invasion by the little red fire ant, *Wasmannia auropunctata*. Biol. J. Linn. Soc. **21**: 229–242.

MacArthur, R.M., and Wilson, E.O. 1967. The theory of island biogeography. Monogr.
Popul. Biol. 1, Princeton Univ. Press, Princeton, N.J.

McMullen, C.K. 1987. Breeding systems of selected Galápagos Island angiosperms.
Am. J. Bot. **74**: 1694–1705.

McMullen, C.K. 1993. Flower-visiting insects of the Galápagos Islands. Pan-Pac.
Entomol. **69**: 95–106.

Peck, S.B. 1996. Diversity and distribution of the orthopteroid insects of the Galápagos
Islands, Ecuador. Can. J. Zool. **74**: 1497–1510.

Peck, S.B., and Forsyth, A. 1982. Composition, structure, and competitive behavior in a
guild of Ecuadorian rainforest dung beetles (Coleoptera: Scarabaeidae). Can. J.
Zool. **60**: 1624–1634.

Peck, S.B., and Kukalova-Peck, J. 1990. Origin and biogeography of the beetles
(Coleoptera) of the Galápagos Archipelago, Ecuador. Can. J. Zool. **68**: 1617–1638.

Peck, S.B., Wigfull, P., and Nishida, G. 1999. Physical correlates of insular species diversity: the insects of the Hawaiian islands. Ann. Entomol. Soc. Am. **92**: 529–536.

Polis, G.A., Hurd, S.D., Jackson, C.T., and Sanchez-Piñero, F. 1997. El Niño effects on
the dynamics and control of an island ecosystem in the Gulf of California. Ecology
78: 1884–1897.

Polis, G.A., Hurd S.D., Jackson, C.T., and Sanchez-Piñero, F. 1998. Multifactor population limitation: variable spatial and temporal control of spiders on Gulf of California Islands. Ecology **79**: 490–502.

Reice, S. R. 1994. Nonequilibrium determinants of biological community structure.
Am. Sci. **82**: 424–435.

Reynolds, R.W., Geist, D., and Kurz, M.D. 1995. Physical volcanology and structural
development of Sierra Negra volcano, Isla Isabela, Galápagos Archipelago. Geol.
Soc. Am. Bull. **107**: 1398–1410.

Schatz, H. 1998. Oribatid mites of the Galápagos Islands—faunistics, ecology and
speciation. Exp. Appl. Acarology **22**: 373–409.

Simpson, B.B. 1974. Glacial migrations of plants: Island biogeographical evidence.
Science **185**: 698–700.

Sousa, W. P. 1984. The role of disturbance in natural communities. Ann. Rev. Ecol.
Syst. **15**: 353–392.

Thornton, I.W.B. 1967. The measurement of isolation on archipelagos, and its relation
to insular faunal size and endemism. Evolution **21**: 842–849.

Thornton, I.W.B. 1971. Darwin's Islands, a natural history of the Galápagos. Natural
History Press, Garden City, NY. 322 pp.

Usinger, R.L. 1972. Robert Leslie Usinger: Autobiography of an entomologist. Mem.
Pacific Coast Entomol. Soc. vol. 4. 330 pp.

Wardle, D.A., Zackrisson, O., Hörnberg,G., and Gallett, C. 1997. The influence of island area on ecosystem properties. Science **277**: 1296–1299.

Wiggins, I.L., and Porter, D.M. 1971. The flora of the Galápagos Islands. Stanford University Press, Stanford, CA.

Chapter 6
Order Collembola: Springtails

Superclass Hexapoda
Class Parainsecta
Order Protura

It is not known if Proturans live in the Galápagos. They are very small, and special collecting techniques are needed to find them in soil and leaf litter. The proturan fauna of the entire Neotropics is virtually completely unstudied and unknown. Because of the diversity of the small soil dwelling Collembola known in the Galápagos, I expect that Protura do occur in the islands and will be discovered eventually.

Order Collembola
The Springtails

7 families, 24 genera, 35 species
(10 endemic?, 23 indigenous?, 2 adventive?)
Minimally 33 natural colonizations?

Collembola, or springtails, are small hexapods that live in concealment and feed as scavengers or decomposers in leaf litter, decaying wood, soil, or on fungi. Surprisingly, many are now known from the Galápagos, but these species mostly occur in low densities, possibly because of the harsh-arid habitats of the lowlands of the islands. Diversity seems to be higher in the moist uplands of the islands. Most collembola are probably not very tolerant of salt water, but some, like *Anurida maritima* certainly live in the littoral zone. It seems most likely that Collembola dispersed to the Galápagos by rafting. The best analysis of the biogeographic complexities of collembola of oceanic islands is that of Christensen and Bellinger (1994) for Hawaii. The Neotropical collembola fauna is not well known, but the recorded fauna of 908 species and 158 genera has been summarized by Mari Mutt and Bellinger (1990). This is estimated to be only 25% of the actual fauna. The classification and distributions used here follow that of Mari Mutt and Bellinger (1990). The Galápagos fauna merits additional study, and the summary given here is only provisional. My own collections have not yet been studied. The specimens are with Dr. H. Palacios-Vargas at UNAM, Mexico City, or the Charles Darwin Research Station.

Suborder Arthropleona
Family Cyphoderidae
Genus *Cyphoderus* Nicolet 1841

C. galapagoensis Jacquemart 1976: 153.
  Distribution. Endemic; Isabela (Volcan Alcedo), Santa Cruz
  Bionomics. Scavenger; littoral, arid, agriculture and humid forest zones;
  leaf litter, rotten *Opuntia*, forest humus at 200 m; January–April.

Family Entomobryidae
Subfamily Entomobryinae
Genus *Lepidocyrtus* Bourlet 1839

L. leleupi Jacquemart 1976: 145.
  Distribution. Endemic; Santa Cruz
  Bionomics. Scavenger; humid forest zone; forest humus at 200 m; November.

Genus *Seira* Lubbock 1870

S. caheni Jacquemart 1976: 149.
  Distribution. Indigenous?; Cuba, Puerto Rico; Marchena, Santa Cruz, San
  Cristóbal
  Bionomics. Scavenger; humid forest zone; forest humus at 200 m; no month
  given.

S. galapagoensis Jacquemart 1976: 151.
  Distribution. Endemic; Pinzón, Santa Cruz, Santiago, San Cristóbal
  Bionomics. Scavenger; humid forest zone; forest humus at 200 m; no month
  given.

Genus *Pseudosinella* Schaeffer 1897

P. intermixta (Folsom) 1924: 75 (*Lepidocyrtus*).
  Distribution. Indigenous?; Peru; Baltra
  Bionomics. Scavenger; arid zone; collected in April by W. Beebe among
  "lepismatid egg shells."

Family Hypogastruridae
Genus *Ceratophysella* Börner 1932

C. denticulata (Bagnall) 1941: 218; Najt et al. 1991: 154.
  Distribution. Probably adventive; cosmopolitan; Mexico, Argentina, Ecua-
  dor; San Cristóbal, Santa Cruz
  Bionomics. Scavenger; arid to pampa zone; in litter; January, February.

Genus *Xenylla* Tullberg 1869

X. yucatana Mills 1938: 183, Da Gama 1986: 272; Najt et al. 1991: 154.
  Distribution. Indigenous?; Pantropical, Mexico, Central America, South
    America, West Indies; Marchena, Pinta, Pinzón, San Cristóbal, Santa
    Cruz
  Bionomics. Scavenger; in all zones; very abundant in humus; January–
    March.

Family Odontellidae
Genus *Xenyllodes* Axelson 1903

X. species. Najt et al. 1991: 156.
  Distribution. Indigenous?; San Cristóbal
  Bionomics. Scavenger; pampa zone; under ferns near El Junco lake; January.

Family Onychiuridae
Subfamily Onychiurinae
Genus *Protaphorura* Absolon 1901

P. encarpata (Denis) 1931: 102; Thibaud et al. 1994: 199.
  P. prosensillata Najt et al. 1991: 163.
  Distribution. Indigenous; USA, Mexico, Costa Rica, Venezuela, Argentina;
    San Cristóbal, Santa Cruz
  Bionomics. Scavenger; in litter in agriculture zone, guava thickets, humid
    forest and pampa zone; January–March.

P. armatus group Najt et al. 1991: 163.
  Distribution. Endemic?; Santa Cruz
  Bionomics. Scavenger; litter between lava blocks in arid zone.

Subfamily Tullbergiinae
Genus *Mesaphorura* Börner 1901

M. yosii (Rusek) 1867: 191; Najt et al. 1991: 165.
  Distribution. Indigenous?; cosmopolitan (Europe, Asia, Pacific Area, North
    and South America); San Cristóbal, Santa Cruz, Santiago
  Bionomics. Scavenger; in litter in agricultural, humid forest, evergreen shrub,
    and pampa zones; January–March.

Family Isotomidae
Subfamily Anurophorinae
Genus *Cryptopygus* Willem 1902

C. thermophilus (Axelson) 1900: 113; Thibaud et al. 1994: 203.

Distribution. Indigenous?; cosmopolitan; Española, Fernandia, Gardner at Floreana, Isabela, Marchena, Pinzón, Santa Cruz, San Cristóbal and Santiago.

Bionomics. Scavenger; arid to pampa and agriculture zones; all vegetation zones, especially abundant in litter in humid forest and pampa zones; January–April.

Genus *Folsomides* Stach 1922

F. parvulus Stach 1922: 17; Thibaud et al. 1994: 200. (as *F. americanus* Denis 1931: 126)

Distribution. Indigenous?; widespread Neotropical; cosmopolitan; Pinzón, Santa Cruz, San Cristóbal.

Bionomics. Scavenger; in humid zone forests, agriculture zone in coffee and avocado groves, in leaf litter and fern litter; January, February.

F. centralis (Denis) 1931: 111; Thibaud et al. 1994: 200.

Distribution. Indigenous?; widespread Neotropical, plus Canaries, Seychelles, southeast Asia and Hawaii; Floreana, Isabela (Tagus Cove), Santa Cruz, San Cristóbal, Santiago.

Bionomics. Scavenger; littoral, arid, agriculture, transition and humid forest zones; in mangrove litter, rotted wood, and leaf litter; January–April.

F. delamarei Thibaud et al. 1994: 200.

Distribution. Endemic; Pinzón, Santa Cruz, San Cristóbal.

Bionomics. Scavenger; arid zone with *Hibiscus* litter; transition and pampa zones; in fern and leaf litter; January, February.

Genus *Folsomina* Denis 1931

F. onychiurina Denis 1931: 128; Thibaud et al. 1994: 200.

Distribution. Indigenous?; USA, Cuba, Mexico to Venezuela and Argentina; also Pacific Islands and Old World; Isabela, Pinta, Santa Cruz, San Cristóbal.

Bionomics. Scavenger; in litter in humid forest and pampa zones; in fern and tree fern litter; in agriculture zone with coffee and avacados; January–March.

Genus *Isotomodes* Axelson 1907

I. denisi Folsom 1932: 59; Thibaud et el. 1994: 200.
  Distribution. Indigenous?; previously known only from Hawaii; Floreana.
  Bionomics. Scavenger; record based on one specimen from guava grove in
    humid forest agriculture zone; April.

I. trisetosus Denis 1923: 62; Thibaud et al. 1994: 200.
  Distribution. Indigenous?; Peru and Argentina; Floreana, Marchena, Pinzón,
    Santa Cruz, Santiago.
  Bionomics. Scavenger; littoral to humid forest zones; in litter and under
    rocks; January–March.

Subfamily Isotominae
Genus *Axelsonia* Börner 1906

A. littoralis (Moniez) 1890: 324; Thibaud et al. 1994: 203.
  Distribution. Indigenous?; Europe and Mediterranean basin, Madagascar
    and Seychelles; Santa Cruz.
  Bionomics. Scavenger; halophile along sea coasts; reported from Santa Cruz
    in littoral and transition zones, under stones; January.

Genus *Isotomiella* Bagnall 1940

I. symetrimucronata Najt and Thibaud 1987: 207; Thibaud et al. 1994: 200.
  Distribution. Indigenous?; Ecuador, Brazil, Seychelles, Fangataufa;
    Floreana, Santa Cruz, San Cristóbal.
  Bionomics. Scavenger; agriculture zone, humid forest, and evergreen shrub
    zones; January–April.

Genus *Isotomurus* Börner 1903

I. opala Christiansen and Bellinger 1992: 155.
  Distribution. Indigenous?; Hawaii; San Cristóbal.
  Bionomics. Scavenger; known from one specimen from under flat stones
    along stream in humid forest; February.

Genus *Rhodanella* Salmon 1945

R. minos (Denis) 1928: 3; Thibaud et al. 1994: 203.
  Distribution. Adventive?; Ivory Coast, Somalia, Zimbabwe; Santa Cruz.
  Bionomics. Scavenger; one specimen from litter in transition zone forest;
    doubtful record; February.

Genus *Psammisotoma* Greenslade & Deharving 1986

P. galapagoensis Thibaud, Najt and Jacquemart 1994: 202.
Distribution. Endemic; Santa Cruz.
Bionomics. Scavenger; littoral zone, under *Sesuvium* and wet soil under *Rhizophora*; arid zone, in litter of *Maytenus* and *Sporobolus*; January.

Family Neanuridae
Subfamily Brachystomellinae
Genus *Brachystomella* Ågren 1903

B. contorta Denis 1931: 80; Najt et al. 1991: 156.
Distribution. Indigenous?; USA to Mexico, West Indies and South America, Africa, Malasia, circumtropical?; Santa Cruz
Bionomics. Scavenger; transition zone litter; January.

B. grottaerti Najt, Thibaud and Jacquemart 1991: 156.
Distribution. Endemic; Marchena, San Cristóbal, Santa Cruz (type locality)
Bionomics. Scavenger; guava forest and pampa zone; January.

Subfamily Frieseinae
Genus *Friesea* Dalle Torre 1895

F. magnicornis Denis 1931: 84; Najt et al. 1991: 157.
Distribution. Indigenous?; Costa Rica and Lesser Antilles; Santa Cruz.
Bionomics. Scavenger; under stones in transition zone; January.

Subfamily Neanurinae
Genus *Americaneura* Cassagnau 1983

A. interrogator Cassagnau and Palacios-Vargas 1983: 7; Najt et al. 1991: 163.
Distribution. Indigenous?; West Indies and South America; Floreana, San Cristóbal, Santa Cruz
Bionomics. Scavenger; in agricultural, humid forest, and evergreen shrub zones; January, February.

Subfamily Pseudachorutinae
Genus *Anurida* Laboulbene 1865

A. marchena Najt, Thibaud and Jacquemart 1991: 161.
Distribution. Endemic; Marchena
Bionomics. Scavenger; on vertical lava wall, with algae, at edge of sea; February.

A. maritima Guerin 1836: 11 (*Achorutes*); Stach 1932: 332
  Distribution. Indigenous?; cosmopolitan; Floreana; to be expected on every island.
  Bionomics. Scavenger; a littoral zone inhabitant, often very abundant in high tide litter deposits; month not given.

Genus *Arlesia* Handschin 1942

A. albipes (Folsom) 1927: 6; Najt et al. 1991: 157.
  Distribution. Adventive; Central and South America, West Indies; Santa Cruz
  Bionomics. Scavenger; in litter of *Hibiscus* and *Croton* in arid zone; January.

Genus *Pseudachorutes* Tullberg 1871

P. difficilis Denis 1931: 88; Najt et al. 1991: 157.
  Distribution. Indigenous?; Costa Rica; Pinta, San Cristóbal, Santa Cruz
  Bionomics. Scavenger; in coffee plantation, and evergreen shrub and humid forest litter; January, February.

P. galapagoensis Najt, Thibaud and Jacquemart 1991: 157.
  Distribution. Endemic; San Cristóbal, Santa Cruz (type locality)
  Bionomics. Scavenger; in litter in humid forest, evergreen shrub and pampa zones; January, February.

P. macrochaetus Najt, Thibaud and Jacquemart 1991: 158.
  Distribution. Endemic; San Cristóbal (type locality)
  Bionomics. Scavenger; in litter in coffee plantation; February.

P. species near *castrii* (Rapoport and Rubio) 1963: 111; Najt et al. 1991: 161.
  Distribution. Indigenous?; *P. castrii* occurs in Mexico; Santa Cruz
  Bionomics. Scavenger; in transition zone forest litter, August.

P. species near *murphyi* Massoud 1965: 384; Najt et al. 1991: 158.
  Distribution. Indigenous?; Santa Cruz (*P. murphyi* is from New Guinea)
  Bionomics. Scavenger; in transition forest litter; August.

Acknowledgements

I thank P. Bellinger (California State University, Northridge, CA) for reviewing this section and the distributions of the species. Dr. H. Palacios-Vargas (UNAM, Mexico City, Mexico) has supplied some bionomic and taxonomic information.

References

Christiansen, K., and Bellinger, P. 1994. The biogeography of Hawaiian Collembola: the simple principles and complex reality. Orient. Insects **28**: 309–351.

da Gama, M.M. 1986. Systématique evolutive des *Xenylla*. XIV. Rev. Suisse Zool. **93**: 271–277.

Folsom, J.W. 1924. Apterygota of the Williams Galápagos Expedition. Zoologica **5**: 67–76. + pls. 3–5.

Jacquemart, S. 1976. XXII. Collembola nouveaux des Îles Galápagos. Mission zoologique belge aux iles Galápagos et en Ecuador (N. et J. Leleup, 1964–1965). **3**: 137–157.

Mari Mutt, J.A., and Bellinger, P.F. 1990. A catalog of the Neotropical Collembola, including Nearctic areas of Mexico. Flora and Fauna Handbook No. 5. Sandhill Press, Gainesville, FL 237 pp.

Najt, J., Thibaud, J.-M., and Jacquemart, S. 1991. Les Collemboles (Insecta) de l'Archipel des Galápagos. I. Poduromorpha. Bull. Inst. Roy. Sci. Nat. Belgique, Entomol. **61**: 149–166.

Stach, J. 1932. The Norwegian Zoological Expedition to the Galápagos Islands 1925, conducted by Alf Wollebaek. III. Die Apterygoten aus den Galápagos-Inseln. Nyt. Magazin Naturvidenskaberne **71**: 331–346, pls. 1–4. Reprinted as Meddeldser fra Zoologiske Museum, Olso **29**: 331–346 (1932).

Thibaud, J.-M., Najt, J., and Jacquemart, S. 1994. Les Collemboles (Insects) de l'Archipel des Galápagos. II. Isotomidae. Bull. Inst. R. Soc. Nat. Belg. Entomol. **64**: 199–204.

Chapter 7
Order Diplura: Bristletails

Class Insecta
Order Diplura
The Bristletails or Diplurans

2 families, 2 genera, 2 species (1 indigenous, 1 adventive)
Minimally 1 natural colonization

Diplura are fragile, elongate, wingless white insects living in damp places in soil, under rocks, and in caves. In the Galápagos they occur mostly in the higher elevations of the islands, and are more often found in the rainy season, when they come up to the soil surface. In the arid zone they are in places with abundant soil moisture, and or in caves and crevices, especially near the water table. They may disperse naturally by rafting or may be transported by humans in soil around the roots of plants. They can run rapidly and must be collected carefully with a moistened brush or aspirator into alcohol. My specimens are at the American Museum of Natural History, New York, NY.

Key to families of Galápagos Diplura

1a   Cerci (at end of abdomen) long, thin, pale coloured, segmented; not
     shaped like forceps . . . . Campodeidae
1b   Cerci short, stout, darkly pigmented, unsegmented; shaped like forceps
     . . . . Japygidae

Family Campodeidae
Genus *Lepidocampa* Oudemans 1890

L. zetekei seclusa Condé 1960: 172;
   L. zeteki Folsom 1927; Paclt 1976: 117; Peck 1990: 368.
   L. juradoi seclusa Condé 1960: 172
   Distribution. Indigenous species and subspecies?; widespread in Neotropics; Santa Cruz (no detailed type locality indicated), Gardner at Española. The species is otherwise reported to occur in Trinidad, Venezuela, Brazil, Argentina and Bolivia (Condé 1960). It is unlikely that this represents an adventive species because of its presence on Gardner at Española.

Bionomics. Scavenger; soil and cave inhabitant; sealevel to 875 m; May–September; littoral and arid to pampa zones. On Santa Cruz known from Cueva del Cascajo, Cuevitas al suroeste de Cerro Banderas, Cueva de Gallardo, Cueva de Kubler, and Cueva Caballo (Hernandez et al. 1992; Peck 1990).

Notes. Paclt (1976) seemed unaware of the record by Condé (1960). Condé and Bareth (1983) placed *L. juradoi* as a synomym of *L. zeteki*. Condé (1976) reported *L. juradoi seclusa* from caves in Trinidad and Venezuela.

Family Japygidae
Genus *Parajapyx* Silvestri 1903

P. isabellae (Grassi) 1886; Paclt 1976: 125; Peck 1990: 368.

Distribution. Adventive; U.S.A., Italy, Chile, Argentina, Panama; Santa Cruz (200 m).

Bionomics. Scavenger; soil inhabitant; probably spread by commerce; many feed on plant roots (Reddell (1985); agricultural zone; July.

Acknowledgements

I thank Lynn Ferguson (Longwood College, Farmville, VA) for help with literature and nomenclature.

References

Condé, B. 1960. Presence des Diploures Campodéides sur les iles Galápagos. Bull. Mus. Nat. d'Hist. Nat. 2nd ser., **32**: 172–176.

Condé, B. 1976. Quelques microarthropodes conservé en Geneva (Palpigrades, Protures, Diploures Campodeidae). Rev. Suisse Zool. **83**: 747–755.

Condé, B., and Bareth, C. 1983. Diploures campodéidés des Petites Antilles. Rev. Suisse Zool. **90**: 689–697.

Hernandez Pacheco, J.J., Izquierdo Zanova, I., and Olmí Masoliver, P. 1992. Resultados Cientificos del Proyecto Galápagos: Patrimonio de la Humanidad. No. 2. Catalogo Espeleologico de las Islas Galápagos. Res. Ci. Entomol. Proy. Galápagos TFMC. 111 pp. Museo Cien. Nat. Tenerife, Canaries.

Paclt, J. 1976. XXI. Diploures et Thysanoures recolte dans les iles Galápagos et en Ecuador par N. et J. Leleup. Mission zoologique belge aux iles Galápagos et en Ecuador (N. et J. Leleup, 1964–1965) **3**: 115–134.

Peck, S.B. 1990. Eyeless arthropods of the Galápagos Islands, Ecuador: composition and origin of the cryptozoic fauna of a young, tropical, oceanic archipelago. Biotropica **22**: 366–381.

Reddell, J.R. 1985. A checklist and bibliography of the Iapygoidea (Insecta; Diplura) of South America. Pearce-Sellards Ser. Tex. Mem. Mus. 42. 34 pp.

Chapter 8
Orders Archeognatha and Thysanura

Jumping Bristletails and Silverfish

3 families, 4 genera, 4 species (2 endemic, 1 indigenous, 1 adventive)
Minimally 3 natural colonizations

These orders are superficially similar, but morphologically very distinct. Both are elongate, brown or white, primitively wingless insects with three cerci. They can jump or run rapidly. They usually hide in the day and emerge at night to crawl on rocks, in litter, and even upon shrubs or cacti. They feed on terrestrial algae, lichens, and litter. *Nicoletia meinerti* is eyeless and occurs in soil or caves. Many archeognathans worldwide live in coastal habitats as well as inland, but on the Galápagos they are known only from upland humid forest or pampa sites. In contrast, Galápagos Thysanura occur in lowland littoral or arid zone habitats. Both groups probably dispersed by rafting. Distributions of the South American taxa are reviewed by Wygodzinsky (1967). See also Wygodzinsky (1959, 1972) for keys to species of Lepismatidae of the Caribbean and northern South America. Some of these species may be eventually found to occur in the Galápagos. My specimens are at the American Museum of Natural History, New York, NY.

Key to the Archeognatha and Thysanura of the Galápagos.

1a Compound eyes large and contiguous; middle and hind coxae with styli (Order Archeognatha; Meinertellidae) . . . . . *Neomachilellus "muticus"*
1b Compound eyes small and widely separated, or absent; middle and hind coxae without styli (Order Thysanura) 2
2a Eyes and ocelli absent, body scales absent (Nicoletiidae)
. . . . *Nicoletia meinerti*
2b Eyes present, body scales present (Lepismatidae). 3
3a Last (tenth) abdominal tergum (pygidium) triangular, sharply pointed; larger, adult body length 10–12 mm; body conspicuously covered with large brownish scales . . . . *Stylifera galapagoensis*
3b Last (tenth) abdominal tergum rounded, not sharply pointed; smaller, adult body length about 5–7 mm; body colour pale or whitish, not as conspicuously entirely covered with large brownish scales
. . . . *Heterolepisma insulare*

Order Archeognatha (= Microcoryphia)
The Jumping Bristletails
Family Meinertellidae
Genus *Neomachilellus* Wygodzinski 1952

N. "muticus" (Banks) 1901: 543 (*Machilis*); Paclt 1976: 126.
Distribution. Indigenous?; Isabela (Cerro Azul, 760–1300 m); Sierra Negra, pampa pit traps, 1000 m); Santa Cruz (Cerro Crocker, Media Luna, Puntudo, Los Gemelos; 450–875 m).
Bionomics. Scavenger; inhabitant of moist leaf litter, under stones, under boards, in sphagnum bogs, and on and under bark; nocturnal; usually taken in pit traps, and by sifting litter; only in upper elevation humid forest to pampa zones; January–August.
Notes. This is a large genus in South America; extending from Argentina through the Caribbean and Mexico to Florida and Georgia. The species are usually restricted to rather moist habitats (Wygodzinsky 1959). Use of the above species name by both Banks and Paclt may be questioned because it is based on material from arid, low lying Clipperton Island (a French possession off the SW coast of Mexico) which does not have the moist forest habitats of Isabela and Santa Cruz. The family has never been found in the littoral or arid zones in the Galápagos. In other parts of the world Archeognatha often occur among rocks on the seashore.

Order Thysanura (= Zygentoma)
The Silverfish
Family Lepismatidae
Genus *Heterolepisma* Escherich 1905

H. insulare (Banks) 1901: 543 (*Lepisma*); Paclt 1959: 171; 1976: 128.
H. intermedia Folsom 1924: 67; Stach 1932: 335.
Distribution. Endemic; Baltra, Bartolomé, Cowley, Darwin, Española, Fernandina, Floreana, Genovésa, Isabela, Pinta, Pinzón, San Cristóbal, Santa Cruz, Santa Fé, Seymour, South Plaza, Tortuga, Wolf.
Bionomics. Scavenger; littoral zone, in litter or soil near beach; arid zone, in litter, under roots of *Bursera*, humus; agricultural zone in litter under coffee trees; rarely up to 600 m on Santa Cruz; January–May.

Genus *Stylifera* Stach 1932

S. galapagoensis (Banks) 1901: 541 (*Lepisma*); Folsom 1924: 71; Stach 1932: 337 (*Acrotelsa*); Paclt 1959: 172; 1976: 130.
Distribution. Endemic; Baltra, Daphne Major, Española, Floreana, Genovésa, Isabela, Marchena, Pinta, Pinzón, Santa Cruz, Santa Fé, Santiago, Seymour

Bionomics. Scavenger; littoral zone, on rock faces or under stones on beach; commonly found running at night on vegetation in arid zone; soil and litter inhabitant; on occasion up to 600 m elevation, January–December.
Notes. This species may be related to or conspecific with *S. gigantea* Escherich, which is widespread in the West Indies, Florida, and coastal Peru (Wygodzinsky 1959, 1967).

Family Nicoletiidae
Genus *Nicoletia* Gervais 1844

N. meinerti Silvestri 1905; Paclt 1976: 130; Peck 1990: 368.
Distribution. Adventive; Neotropical (also west Africa, Marquessas and Hawaiian islands); Isabela, Santa Cruz.
Bionomics. An eyeless white inhabitant of soil and litter in the arid and littoral zones. Paclt (1976) states this to be a modern introduction. Taken in the crevice cave called Grieta Iguana(at CDRS) on Isla Santa Cruz, and rotted logs in a coffee plantation (380 m) at Santo Tomas on Isabela. Some populations may be facultatively parthenogenetic, which will facilitate dispersal; March, December.

References

Banks, N. 1901. Papers from the Hopkins Stanford Galápagos Expedition, 1898–1899. V. Entomological Results (5): Thysanura and Termitidae. Proc. Wash. Acad. Sci. **3**: 541–546.

Folsom, J.W. 1924. Apterygota of the Williams Galápagos Expedition. Zoologica **5**: 67–76.

Palct, J. 1956. Biologie der primär flügellosen Insekten. Jena, Germany: Gustav Fischer Verlag. 258 pp.

Palct, J. 1959. Uber die Lepismatidae (Ins. Thysanura) von den Galápagos-Inseln. Senckenbergiana Biologie **40**: 171–172.

Palct, J. 1976. XXI. Diploures et thysanoures recolte dans les iles Galapagoes et en Ecuador par N. et J. Leleup. Mission zoologique belge aux iles Galápagos et en Ecuador (N. et J. Leleup, 1964–1965) **3**: 115–134.

Peck, S.B. 1990. Eyeless arthropods of the Galápagos Islands, Ecuador: composition and origin of the cryptozoic fauna of a young, tropical, oceanic archipelago. Biotropica **22**: 366–381.

Stach, J. 1932. The Norwegian Zoological Expedition to the Galápagos Islands 1925, conducted by Alf Wolleback. III. Die Apterygoten aus den Galápagos-Inseln. Nyt Magazin Naturvidenskaberne **71**: 331–346, pls. 1–4. [Reprinted as Meddelelser fra det Zoologiske Museum, Oslo, nr. 29, pp. 331–346, pls. 1–4 (1931).]

Wygodzinsky, P. 1959. Thysanura and Machilidae of the Lesser Antilles and northern South America. Stud. Fauna Curaçao Carib. Islands **9**: 28–49.

Wygodzinsky, P. 1967. On the geographical distribution of the South American Microcoryphia and Thysanura (Insecta). Biol. de l'Amerique Australe **3**: 505–524.

Wygodzinsky, P. 1972. A revision of the silverfish (Lepismatidae, Thysanura) of the United States and the Caribbean area. Am. Mus. Novit. **2481**: 1–26.

Chapter 9
Order Odonata

The Dragonflies and Damselflies

3 families, 8 genera, 9 species (1 endemic, 8 indigenous)
Minimally 8 natural colonizations

Although there is a very rich odonate fauna in mainland South America, that of the Galápagos is impoverished. For instance, Venezuela is known to have 449 species, 116 genera, and 14 families of Odonata. No such list seems to have been prepared for mainland Ecuador. The indigenous Galápagos species all have very large New World distributions, and the fauna is clearly derived from continental tropical America (Schmidt 1938). All dispersal to the Galápagos was by active flight, perhaps aided by wind. The 900–1000 km separating the islands and the mainland is not likely to be much of an effective barrier for wandering or migratory Odonata species. Since long distance migration swarms of Odonata have been found far at sea (Russell et al. 1998), their presence on the islands is not surprising. But the very limited number of species actually in the Galápagos may be a surprise. It is probably a reflection of the limited fresh water available for nymphal development. It is the seasonal and harsh aridity of the islands which must have limited colonization to that of the few known species. The highlands of most of the large islands may have seasonal streams or pools for odonate reproduction, but these can completely vanish in a dry season. I have seen that nymphs of some species may pass the dry season in the highlands by burrowing into the mud, but if an island population does not survive a dry season, then recolonization of the island is necessary. Permanent fresh water occurs only on San Cristóbal Island at El Junco Lake, its outlet stream, and the reservoir and stream at La Toma. This island (and possibly Isabela and Santa Cruz, from which adults but no nymphs are known) supports the only endemic dragonfly.

Adult Odonata are often found in the Galápagos flying about any temporary body of fresh water in the highlands of the islands, or around coastal mangrove lagoons where they feed on flying insects. Their predatory nymphs are seen in water pools in temporary streams near Media Luna, Santa Cruz, and in submerged grasses or sedges at the edges of apparently fresh water coastal lagoons (or in fresh water lenses floating above the salt water) near Villamil, Isabela. Most odonate nymphs cannot tolerate saline conditions so they would not be expected in saline coastal waters, but some do apparently live in brackish water. Biological data on the species in places other than the Galápagos can be found in Needham and Westfall (1955), Needham et al. (2000), and Dunkle (1989, 1990). Corbet (1962, 1980) provides general reviews of Odonata biology. The body colors noted in the key are best observed in live specimens. They fade after death. Keys to genera of adults and nymphs are in Merritt and Cummins (1984).

No detailed or quantative ecological studies exist for Galápagos Odonata. Little attempt has been made to document the presence or absence of nymphs in various water-bodies. Unidentified nymphs have been reported in the volcanic crater lake at the summit of the active volcano of Isla Fernandina (Eibl-Eibesfeldt 1961: 102) so they could also occur in the summit crater-lake on the volcano of Cerro Azul, Isla Isabela, or in lakes in other volcano craters. During and following the unusually rainy conditions of the El Niño of January–May 1992, I found hawking dragonflies to be exceptionally abundant along the coasts of many islands. The ecology of the Galápagos limnic fauna was reviewed by Gerecke et al. (1995) and related to that of other oceanic islands by Peck (1992).

Suborder Zygoptera
The Damselflies
Family Coenagrionidae

Genus *Ischnura* Charpentier 1840

I. hastatum (Say) 1839: 38: (*Agrion*); Currie 1901: 382; Calvert 1906: 130; Gloger 1964: 3; Turner 1967: 285, 290 (*Anomalagrion*); Peck 1992: 311.
   Distribution. Indigenous; United States, Caribbean, Neotropics; Isabela (abundant near Villamil), San Cristóbal (abundant around El Junco), Santiago, Santa Cruz (abundant near Media Luna)
   Bionomics. Littoral saline meadows to pampa zones, around permanent wet spots and pools; January–May, December.

Suborder Anisoptera
The Dragonflies

A. Key to Adult Galápagos Anisoptera

1a   Triangles (t) in front and hind wings similar in shape and about equidistant from arculus (ar) (Fig. 9.1); brace vein below proximal end of stigma; hind wing without foot-shaped anal loop (al), "toe" absent (Fig. 9.1) Aeshnidae 2
1b   Triangles (t) in front and hind wings not similar in shape and not equidistant from arculus (Fig. 9.2); no brace vein below proximal end of

Figs. 9.1–9.6. Characters of Galápagos dragonflies. Fig. 9.1. Wing base of *Aeshna galapagoensis*, showing similar triangles (t) of Aeshnidae; sv. sector veins; ar, arculus; al, anal loop. Fig. 9.2. Wings of *Pantala hymenaea*; M2, undulated M2 vein. Fig. 9.3. Hind wing of *Brachymesia herbida* showing two paranal cells before start of anal loop. Fig. 9.4. Hind wing of *Tramea cophysa* showing three paranal cells (pc) before start of anal loop (al). Fig. 9.5. Male abdominal tip of *T. calverti*. Fig. 9.6. Male abdominal tip of *T. cophysa*. Figs. 9.5 and 9.6 from De Marmels and Racenis 1982.

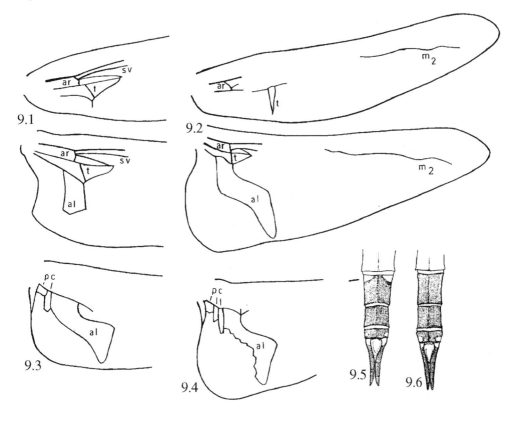

stigma; hind wing with foot shaped anal loop (al), with well developed
toe (Fig. 9.2) Libellulidae. 3
2a   Sector veins (sv) arising from arculus vein near its middle (Fig. 9.1);
     thorax multicoloured; male with anal border of hind wing angulate; au-
     ricles (lateral earlike projections) on male 2nd abdominal segment
     . . . . *Aeschna galapagoensis*
2b   Sector veins arising from arculus vein in its upper half; thorax green;
     anal border of hind wing rounded in both sexes; auricles absent on
     male 2nd abdominal segment. . . . . *Anax amazili*

3a Wing vein M2 smoothly curved 4

3b Wing vein M2 waved or undulating (Fig. 9.2); *Pantala* 5

4a Hind wing with 2 paranal cells (pc) (behind anal vein) before the start of the anal loop (al) (Fig. 9.3); no brown band at hind wing base . . . . *Brachymesia herbida*

4b Hind wing with 3 paranal cells (pc) before start of anal loop (al) (Fig. 9.4); with brown band at hind wing base 6

5a Hind wings with large brown spot at base near broadly rounded anal angle . . . . *Pantala hymenaea*

5b Hind wing with no spot near base or on anal angle. . . *Pantala flavescens*

6a Wings with pterostigma trapezoidal, front side distinctly longer than rear; some double-length cells above apical planate cell, reaching from planate to M1. *Tramea* 7

6b Wings with front and rear sides of pterostigma about equal in length; apical planate cell poorly developed and with no double length cells above it . . . . *Erythemis vesiculosa*

7a Underside of abdomen brown to red; abdominal segment 8 with a semi-circular pale basal-lateral spot (Fig. 9.5); hindwing tinged brown with an amber-edged basal band; male with only broad band on top of frons and face otherwise pale; male at maturity with lower frons and face red . . . . *Tramea calverti*

7b Underside of abdomen black; abdominal segment 8 all black (Fig. 9.6); hindwing clear with sharply edged dark basal band; male at maturity with frons all violet, male face black . . . . *Tramea cophysa*

B. Key to nymphs and nymphal exuviae of Galápagos Anisoptera
by Nancy L. House (Carleton University, Ottawa, Canada),
excluding *Erythemis*

1a. Prementum and labial palps flat (Aeshnidae). 2

1b. Prementum and labial palps spoon-shaped (Libellulidae) 3

2a. Compound eyes as long as their greatest width; lateral spines on abdominal segments 7–9 . . . . *Anax amazili*

2b Compound eyes shorter than their greatest width; lateral spines on abdominal segments 6 or 7 . . . . *Aeshna galapagoensis*

3a Dorsal hooks present on abdominal segments; epiproct as long as paraprocts 4

3b Dorsal hooks absent on abdominal segments; length of epiprocts and paraprocts variable 5

4a Dorsal hooks on abdominal segments 3–10; dorsal hook on segment 10 very small while dorsal hook on segment 9 extends well past cerci . . . . *Brachymesia herbida*

4b Dorsal hooks on abdominal segments 3?–8 reduced to knobs with tufts of setae; lateral spines on abdominal segment 9 short, barely extending past abdominal segment 10; labial palp crenations so low as to be almost straight; species not known as adults, ... possibly *Macrodiplax*

5a Epiproct shorter than paraprocts; lateral spines on abdominal segment 8 slightly shorter than the lateral spines on segment 9; lateral spines on segment 9 extending beyond tips of cerci; movable hook of labial palp long and slender . . . . *Tramea calverti* or *Tramea cophysa*

5b Epiproct as long or longer than paraprocts; lateral spines of abdominal segment 8 approximately one third as long as those of segment 9; lateral spines of abdominal segment 9 not extending as far as tips of cerci; movable hook of labial palps short and robust (*Pantala*) 6

6a Base of lateral spine on abdominal segment 9 greater than one third length of spine; epiproct pointed abruptly and curved slightly downwards . . . . *P. hymenaea*

6b Base of lateral spine on abdominal segment 9 less than or equal to length of spine; epiproct gradually pointed and straight . . *P. flavescens*

Superfamily Aeshnoidea
Family Aeshnidae
The Darners
Genus *Aeshna* Fabricius 1775

A. galapagoensis Currie 1901: 382; Martin 1908: 50; Needham 1902: 695; Calvert 1952: 258 (*Neureclipta*); 1956: 137; Asahina 1961: 1; Gloger 1964: 4; Turner 1967: 288, 290; Peck 1992: 314.

Distribution. Endemic; Isabela, San Cristóbal (La Toma), Santa Cruz

Bionomics. Arid to pampa zones; January, March–May, December. A nymph has been taken in a brackish pool at the bottom of Grieta Iguana (a cave) at CDRS.

Notes: Schmidt (1938) states that this species is near *A. corrigera* and *A. californica*, both of California. The endemic status of this species might merit reexamination.

Genus *Anax* Leach 1815

A. amazili (Burmeister) 1839: 841 (*Aeschna*); Asahina 1961: 1; Gloger 1964: 4; Turner 1967: 288, 290; Peck 1992: 314.

Distribution. Indigenous; southern USA (vagrant), Mexico and West Indies, south to Argentina; Bartolomé, Floreana, San Cristóbal, Santa Cruz, Santiago

Bionomics. Arid to pampa zone; January, May. It breeds in weedy ponds, and can breed in brackish water. Nymphs have been collected in brackish lagoons at Villamil, Isla Isabela.

Superfamily Libelluloidea
Family Libellulidae
The Common Skimmers
Genus *Brachymesia* Kirby 1889

B. herbida (Gundlach) 1889: 261 (*Libellula*); Peck 1992: 314.
  Cannacria batesii Kirby 1889: 341.
  Cannacria fumipennis Currie 1901: 387; Calvert 1908: 326; Turner 1967: 290.
  Distribution. Indigenous; Texas to West Indies, south to Argentina; Isabela, Santa Cruz.
  Bionomics. Arid and littoral zones; January–March. It breeds in ponds and marshes, and nymphs can live in brackish water. Although there have been no literature records of new collections of the species since 1901, it was the most abundant species in my 1989 collections.

Genus *Erythemis* Hagen 1861

E. vesiculosa (Fabricius) 1775: 421 (*Libellula*); Needham et al. 2000: 664. The great pondhawk.
  Distribution. Indigenous; southern USA, Mexico, Cuba, Hispaniola, Jamaica, Puerto Rico, south to Paraguay; Isabela.
  Bionomics. Arid zone; May; record based on one adult male taken in 1992 at lagoons west of Villamil.

Genus *Macrodiplax* Brauer 1868

M. balteata Hagen 1861; Gerecke et al. 1995: 132.
  Distribution. Indigenous?; southern USA and Greater Antilles; Isabela.
  Bionomics. Record based on exiuvium of one nymph from lagoons west of Villamil; littoral-arid zone; January. Needham and Westfall (1955) report that this small species is the only member of the genus in the New World. Confirmation of the record is needed to learn if it is established in the Galápagos.

Genus *Pantala* Hagen 1861

P. flavescens (Fabricius) 1798: 285 (*Libellula*); Currie 1901: 385; Calvert 1908: 307; 1947b: 227; Turner 1967: 289; Peck 1992: 315.
  Distribution. Indigenous; cosmopolitan, including Hawaii, but not breeding in Europe; Baltra, Floreana, Isabela, San Cristóbal (La Toma), Santa Cruz, Santiago.
  Bionomics. Littoral and arid to pampa zones; March–May. It is often taken on vessels far at sea. The species breeds in open temporary ponds, including brackish waters. It drifts with the wind and feeds on insects in the aerial plankton.

P. hymenaea (Say) 1839: 18 (*Libellula*); McLachlan 1877: 84; Currie 1901: 385; Calvert 1908: 309; Gloger 1964: 5; Turner 1967: 289; Peck 1992: 315.

Distribution. Indigenous; North, Central and South America and the West Indies; Española, Floreana, Isabela, Santa Cruz, Santiago.

Bionomics. Littoral to pampa zone; March, April, June–October. The species breeds in open temporary pools, including brackish waters. Nymphal development may be as quick as 5 weeks. Nymphs have been found in temporary pools on Santa Cruz.

Genus *Tramea* Hagen 1861

T. calverti Muttkowski 1910; Peck 1992: 315.

Distribution. Indigenous; a vagrant in Florida, USA, and resident in southern Texas, Mexico, the West Indies, Central and South America to Argentina; supposedly Floreana, Santa Cruz. No recent records; those in Peck (1992) belong to the next species.

Bionomics. Littoral and arid zones; January.

T. cophysa darwini Calvert 1947*a*: 227; Asahina 1961: 2; Turner 1967: 288.

T. darwini Kirby 1889: 315 (type locality; "Galápagos Islands"); Currie 1901: 386.

Distribution. Indigenous; in South America from Venezuela through Ecuador to Argentina; Española, Floreana, Isabela, San Cristóbal, Santa Cruz, Santiago.

Bionomics. Arid to pampa zones; December–August. Breeds in temporary or permanent pools and quiet waters, and probably in brackish water (Dunkle 1989). Nymphs have been collected in brackish lagoons at Villamil, Isla Isabela, and temporary ponds on Isla Santa Cruz.

Notes: This species may have been confused with *T. calverti*, and Galápagos literature records of *T. calverti* may really refer to *T. cophysa*. De Marmels and Racenis (1982) clarify the characters and distributions of *T. calverti* and *T. cophysa*, and list only *T. cophysa* from the Galápagos. The key in De Marmels and Racenis (1982) should be consulted. Dunkle (1989, and personal communication) has seen material of *T. calverti*, but not *T. cophysa*, from the Galápagos. My material contains only specimens of *T. cophysa*. If *T. calverti* was actually once present and is now absent in the Galápagos it represents one of the rare cases of a natural extirpation (local extinction) of a population in the archipelago. The species are best separated by the colour of live individuals, as follows: *T. cophysa*; underside of abdomen black, abdominal segment 8 all black, hindwings clear with sharply edged dark basal band, male with frons all violet, male face black at maturity. *T. calverti*; underside of abdomen brown to red, segment 8 with a semicircular pale baso-lateral spot, hindwing tinged brown with an amber-edged band, male with only top of frons violet, male at maturity with lower frons and face red. The questionable species record of *T. calverti* is not used in faunal totals or analyses.

Acknowledgements

Sid Dunkle (Collin County Community College, Plano, TX) provided help with Odonata nomenclature and literature. Oliver Flint (USNM) provided the identification for *Erythemis vesiculosa*; and he and N. Donnelly helped confirm my specimen records as *T. cophysa*.

References

Asahina, S. 1961. Dragonflies taken by Dr. Sekiguchi in the Galápagos Islands. Publication of the Entomological Laboratory, College of Agriculture, University of Osaka Sakai, no. 6, pp. 1–3, 1 figure.

Calvert, P.P. 1908. Insecta. Neuroptera. Odonata. Biologia Centrali-Americana 30: 17–419, pls, 1 map.

Calvert, P.P. 1947*a*. The Odonata collections of the California Academy of Sciences from Baja California and Tepic, Mexico, of 1889–1894. Proc. Calif. Acad. Sci. ser. 4, **23**: 603–609.

Calvert, P.P. 1947*b*. Odonata of voyages under the auspices of the New York Zoological Society. Entomol. News **58**: 227–230.

Calvert, P.P. 1952. New taxonomic entities in Neotropical *Aeshna* (Odonata: Aeshnidae). Entomol. News **63**: 253–264.

Calvert, P.P. 1956. The Neotropical species of the "subgenus *Aeschna*" *sensu* Selys 1883 (Odonata). Mem. Am. Entomol. Soc. **15**: 1–251; 47 pls., 19 tables, maps I–IV.

Corbet, P.S. 1962. A biology of dragonflies. H.F., and G. Witherby Ltd., London. 247 pp.

Corbet, P.S. 1980. Biology of Odonata. Ann. Rev. Entomol. **25**: 189–217.

Currie, R.P. 1901. Papers from the Hopkins Stanford Galápagos Expedition, 1898–1899. Entomological results (3): Odonata. Proc. Wash. Acad. Sci. **3**: 381–389.

De Marmels, J., and Rácenis, J. 1982. An analysis of the *Cophysa*-group of *Tramea* Hagen, with descriptions of two new species (Anisoptera: Libellulidae). Odonatologica **11**: 109–128.

Dunkle, S.W. 1989. Dragonflies of the Florida peninsula, Bermuda and the Bahamas. Scientific Publishers, Gainesville, Florida. 154 pp.

Dunkle, S.W. 1990. Damselflies of Florida, Bermuda and the Bahamas. Scientific Publishers, Gainesville, Florida. 148 pp.

Eibl-Eibesfeldt, I. 1961. Galápagos, the Noah's ark of the Pacific. Doubleday and Co., Garden City, N.Y. 192 pp.

Gerecke, R., Peck, S.B., and Pehofer, H.E. 1995. The invertebrate fauna of the island waters of the Galápagos Archipelago (Ecuador): a limnological and zoogeographical summary. Arch. Hydrobiol. Suppl. **107**: 113–147.

Gloger, H. 1964. Bemerkungen uber die Odonaten-Fauna der Galápagos-Inselin nach der Ausbeute von Juan Foerster, 1959. Opusc. Zool. (Muenchen) **74**: 1–6.

Kirby, W.F. 1889. A revision of the subfamily Libellulinae, with descriptions of new genera and species. Trans. Zool. Soc. London **12**: 249–348, pls. 51–57.

Kirby, W.F. 1890. A synonymic catalogue of Neuroptera Odonata or dragonflies with an appendix of fossil species. London. pp. 1–202.

Martin, R. 1908. Collections zoologiques Edm. de Selys Longchamps. Fasc. 18, Aeschnines.

Merritt, R.W. and Cummins, K.W. 1984. An introduction to the aquatic insects of North America, 2nd ed. Kendall/Hunt Publ., Dubuque, Iowa. 722 pp.

McLachlan, R. 1877. IX. Neuroptera. *In* Account of the zoological collection made during the visit of H.M.S. 'Peterel' to the Galápagos Islands. *Edited by* A. Gunther. Proc. Zool. Soc. London 1877: 84–86.

Needham, J.G. 1902 (1904). New dragon-fly nymphs in the United States National Museum. Proc. U. S. Natl. Mus. **27**: 685–720, 7 pls.

Needham, J.G., and Westfall, M.J., Jr. 1955. A manual of the Dragonflies of North America (Anisoptera) including the Greater Antilles and the provinces of the Mexican Border. Univ. California Press, Berkeley, CA. 615 pp.

Needham, J.G., Westfall, M.J., Jr., and May, M.L. 2000. Dragonflies of North America, revised edition. Scientific Publ., Washington, D.C. 939 pp.

Peck, S.B. 1992. The dragonflies and damselflies of the Galápagos Islands, Ecuador (Insects: Odonata). Psyche (J. Entomol.) **99**: 309–320.

Ris, F. 1913. Libeluline. Collections zoologiques du Baron Edm. de Selys Longchamps, Fasc. 15, pp. 837–964, pl. 7; 16, pp. 965–1042, pl. 8.

Russell, R.W., May, M.L., Soltesz, K.L., and Fitzpatrick, J.W. 1998. Massive swarm migrations of dragonflies (Odonata) in eastern North America. Am. Midl. Nat. **140**: 325–342.

Schmidt, E. 1938. Check-list of Odonata of Oceania. Ann. Entomol. Soc. Am. **31**: 322–344.

Turner, P.E., Jr. 1967. Odonata of the Galápagos Islands. Pan-Pac. Entomol. **43**: 285–291, Figs. a–j, 1 table.

Chapter 10
Order Orthoptera

The Grasshoppers, Katydids and Crickets

3 families, 16 genera (2 endemic), 33 species
(28 endemic, 1 indigenous, 4 adventive)
Minimally 16 natural colonizations

The Order Orthoptera contains a varied assemblage of insects, which are usually herbivores, but some may be predators or scavengers. The systematics of the island grasshoppers has attracted much attention and many species and sub-species names were once proposed which are not now accepted. Considering the richness of the mainland Ecuador fauna, the island fauna is highly impover-ished. Colonization may have been by flight (and wind) for strong fliers such as *Schistocerca*, *Sphingonotus*, and *Neoconocephalus*. Rafting is more likely for weak fliers and flightless groups, especially in the Gryllidae (Nemobiinae and Mogoplistinae). Seventy–four percent of the endemic species are flightless. Loss of flight ability on the Galápagos has occurred in *Halmenus* (maybe it is a descendent from a *Schistocerca* locust), *Closteridea*, *Gryllus,* and *Conocephalus*. Of special note is the evolution of the two species of eyeless sub-terranean *Anurogryllus* crickets. No eyed ancestor of these is known in the is-lands. Evolution of endemics in the Orthoptera has been relatively frequent when compared to most other insect groups.

Many male crickets and katydids (long–horned grasshoppers) stridulate or "sing" at night. There may be different songs for territorial defense, for alarm, or for calling a mate. Females recognize and respond to a song from a male, but only if he is of her species, and if she is at the same temperature. Sometimes the calling songs are more characteristic of the species than are any morphological differences. When two or more species of the same genus live together (in sympatry) they may be most easily separated by their songs. Song variation be-tween separate island populations could reflect either within–species "dialects" or real between–species differences. Laboratory rearings would be needed to separate these possibilities. Songs of Darwin's finches have been studied in de-tail (Grant 1986), but not the songs of most Galápagos Orthoptera. Rich research potential may be present. Virtually hundreds of "song species" of crickets have been found in Hawaii (Otte 1989). This behaviour has been studied only in part in the Galápagos (Otte and Peck 1997, 1998*b*). Single island species of Galáp-agos *Gryllus* seem to have much more variation in their song than is normally found in them elsewhere. It may mean that selection for song consistency has been relaxed in the islands.

The Galápagos fauna fits generalizations for other oceanic island Orthoptera. There is reduced diversity of higher taxa, and a high proportion of both endemic and adventive species (Peck 1996).

The principal references for the Galápagos Islands Orthoptera are Butler (1877), Scudder (1893), McNeill (1901), Snodgrass (1902), Hebard (1920), Caudell (1932), Dirsch (1969), and Otte and Peck (1998*a*, *b*). Peck (1996) is a previous summary of the fauna. Otte and Peck (1997) describe an interesting group of seven endemic species of *Gryllus*. The higher classification of Orthoptera is not widely agreed upon. I here use the conservative system of Borror et al. (1989).

Voucher specimens from my collection are in the Academy of Natural Sciences at Philadelphia, the California Academy of Sciences (*Cycloptilum*), and the National Museum of Natural History, Washington.

Key to the adults of the superfamilies of Galápagos Orthoptera

1a Antennae short, with less than 30 segments, half of body length or less
. . . . . . suborder Caelifera, superfamily Acridoidea, family Acrididae
1b Antennae long, with well over 30 segments, as long as body or longer
(suborder Ensifera; the crickets and katydids) 2
2a All tarsi 4–segmented; vertex of head pointed; ovipositor sword–
shaped
. . . superfamily Tettigonioidae; long–horned grasshoppers and katydids
2b All tarsi 3–segmented; vertex of head rounded; ovipositor needle
shaped. . . . . superfamily Grylloidea; the crickets

Suborder Caelifera
Superfamily Acridoidea
Family Acrididae
The Short–horned Grasshoppers

Key to Galápagos subfamilies of Acrididae

1a Prosternum without median spine. . . . . Oedipodinae
1b Prosternum with median spine . . Cyrtacanthacridinae and Catantopinae

Subfamily Oedipodinae
The Band-winged Grasshoppers

Key to Galápagos species of Oedipodinae

1a Tegmina vestigial (flightless), colour rust brown, mottled with dark-brown spots; outer and inner sides of hind femur with two transverse brown bands; relatively uncommon. . . . . *Closteridea bauri*
1b Tegmina fully developed, grayish, with two transverse brown bands; hind femur without transverse brown bands; common and widespread
. . . . *Sphingonotus fuscoirroratus*

Genus *Closteridea* Scudder 1893 (endemic genus)

C. bauri Scudder 1893: 9; Hebard 1920: 319; Dirsch 1969: 44; Peck 1996: 1502.
  Distribution. Endemic; San Cristóbal (type locality; Wreck Bay), Gardner at Floreana; an old Isabela record from Cowley Mountain (= Volcan Alcedo) is probably in error
  Bionomics. Herbivore; arid to pampa zone; February, August. This is a flightless species which probably lost its wings on the Galápagos.

Genus *Sphingonotus* Fieber 1853

S. fuscoirroratus (Stål) 1860: 345 (*Oedipoda*); McNeill 1901: 495; Caudell 1931: 4; Dirsch 1969: 47; Peck 1996: 1502.
  S. tetranesiotis Snodgrass 1902: 444
  S. t. tetranesiotis Snodgrass 1902: 444 (as variety *charlesensis*; type locality, Floreana)
  S. t. barringtonensis Snodgrass 1902: 445 (as variety; type locality, Santa Fé)
  S.t. hoodensis Snodgrass 1902: 446 (as variety; type locality, Española)
  S. t. indefatigabilensis Snodgrass 1902: 447 (as variety; type locality, Santa Cruz)
  S. trinesiotis Snodgrass 1902: 439
  S. t. trinesiotis Snodgrass 1902: 439 (as variety *chathamensis*; type locality, San Cristóbal)
  S. t. indefatigabilensis Snodgrass 1902: 441 (as variety; type locality, Santa Cruz)
  S. t. albemarlensis Snodgrass 1902: 443 (as variety; type locality, Isabela)
  Distribution. Endemic (or indigenous; the species may also occur in Isla Puna, in the Gulf of Guayaquil); Baltra, Bartolomé, Española, Fernandina, Floreana, Gardner at Española, Isabela, Pinta, Rábida, Santa Fé, San Cristóbal, Santa Cruz, Santiago, Seymour

Bionomics. Herbivore; arid to pampa zone; frequent on open sand, soil, or cinders; sometimes at uv lights; January–June.

Notes. Dirsch (1969) notes the unusual distribution of the genus in the whole of Europe, Africa, most of Asia, and Australia, but in the New World only in Jamaica, mainland Ecuador, and the Galápagos. Dirsch (1969) justifies placing all the previously proposed names as synonyms of a single species by saying that they are not separable by either external morphology or the phallic complex. But remarkable colour differences do exist between populations on the same and different islands. The significance or origin of these "colour forms" should be studied.

Subfamily Cyrtacanthacridinae
The Spur-throated Grasshoppers

Key to Galápagos species of Crytacanthacridinae (and Catantopinae)

1a Adult tegmina and flight wings as long as abdomen or longer 2
1b Adult tegmina only 1/3 length of abdomen, flight wings reduced; *Halmenus* 4
2a Body small, 13.5 mm long; pronotum dorsally with three distinct dark longitudinal stripes separated by two pale stripes (known only from the type specimen) (Catantopinae) . . . . *"Desmopleura concinna"*
2b Body large, 27–65 mm long; *Schistocerca* 3
3a Body large; males about 40 mm long, females about 54 mm long; tegmina pale in basal anterior 1/3 and darker in remaining 2/3; on all islands except Española . . . . *Schistocerca melanocera*
3b Body smaller; males about 31 mm long, females about 40 mm long; tegmina background pale with distinct spots; only on the eastern and southern islands (Genovesa, San Cristóbal, Española, Floreana) . . . . *Schistocerca literosa*
4a Tegmina overlapping dorsally; anterior margin of tegmen strongly convex, apex narrowed; Islas Rábida, Santa Cruz, Santiago . . . . *Halmenus robustus*
4b Tegmina not overlapping dorsally; anterior margin of tegmen straight or only slightly convex, apex rounded or slightly attenuate 5
5a Male cerci narrow and angular, apex obtuse or acute; interocular space narrower than or as wide as antennal scape; upper outer margin of upper valve of ovipositor irregularly serrated 6
5b Male cerci wide and angular, apex obtuse, almost rounded; interocular space wider than antennal space; upper margin of upper valve of ovipositor smooth; Isla Wolf . . . . *Halmenus eschatus*

6a Male cerci narrow, at apex acute or subacute; tegmina reaching or slightly exceeding middle of second abdominal tergite; Isla Isabela
. . . . *Halmenus cuspidatus*
6b Male cerci narrow and comparatively short, apex obtuse; tegmina slightly exceeding end of first abdominal tergite; Isla Floreana
. . . . *Halmenus choristopterus*

Genus *Desmopleura* Scudder 1893

D. concinna Scudder 1893: 19; Dirsh 1969: 43; Peck 1996: 1504.
    Distribution. An error; Santiago (= San Salvador) (erroneous type locality). Known only from the type specimen. This is probably a mislabelled specimen of *Dichroplus vittiger* of Argentina (M. Cigliano, pers. comm.).
    Bionomics. The fact that this winged grasshopper has never been found again on the Galápagos suggests that the record was based on an incorrectly labeled specimen. It should not be considered a member of the island fauna.

Genus *Halmenus* Scudder 1893 (endemic genus)

Adults in this genus do not develop fully formed tegmina or flight wings, so adults are flightless and superficially appear to be nymphs. This and *Schistocerca* are the only New World members of this otherwise entirely Old World subfamily. *Halmenus* may have developed from a very early colonization of a *Schistocerca*-like ancestor (Dirsch 1974), but an analysis of their cuticular hydrocarbons does not support this idea (Chapman et al. 2000).

H. choristopterus Snodgrass 1902: 450; Dirsh 1969: 41; Peck 1996: 1504. Type locality: Floreana.
    H. robustus choristopterus Snodgrass, Hebard 1920: 328
    Distribution. Endemic; Floreana and Gardner at Floreana
    Bionomics. Herbivore; arid to humid forest zones; February–June.

H. cuspidatus Snodgrass 1902: 450; Dirsh 1969: 39; Peck 1996: 1504. Type locality: Iguana Cove, Isabela.
    H. robustus cuspidatus Snodgrass, Hebard 1920: 328
    Distribution. Endemic; Isabela (Cerro Azul, Sierra Negra, Volcán Darwin, Volcán Alcedo)
    Bionomics. Herbivore; arid to pampa zone; January–December.

H. eschatus Hebard 1920: 329; Dirsh 1969: 42; Peck 1996: 1504. Type locality: Wolf.
    Distribution. Endemic; Wolf (type from mockingbird stomach)
    Bionomics. Herbivore; arid zone; May and September.

Remarks. This species seems to be known only from the type and one other specimen.

H. robustus Scudder 1893; 18; McNeill 1901: 496; Caudell 1931: 4; Dirsh 1969; 39; Peck 1996: 1504.

> *H. robustus robustus* Scudder, Hebard 1920: 327; Type locality: Conway Bay, Santa Cruz.
> Distribution. Endemic; Genovesa (Caudell 1931: 4; a record which need verification), Rábida, Santa Cruz, Santiago
> Bionomics. Herbivore; arid to pampa zone; February–July

Genus *Schistocerca* Stål 1873

This is a relatively large genus with species through much of the northern Neotopical Realm and northwards through much of the USA. Otte (1989) suggests that the two Galápagos species are derived from ancestors that colonized by migratory flights across the Pacific from South America. The migratory plague locust of Africa probably represents another colonization from South America, across the Atlantic. The Galápagos species are supposedly not closely related to any mainland species (Dirsch 1974: 1), but an analysis of their cuticular hydrocarbons shows the Galápagos species to be more similar to each other and to differ from the other species which have been studied (Chapman et al. 2000).

S. literosa (Walker) 1870: 620 (*Acridium*); Butler 1877: 88 (*Acridium*) Hebard 1920: 326; Caudell 1931: 4; Dirsh 1969: 35; 1974: 122; Peck 1996: 1503.

> *S. discoidalis* Scudder 1893: 16 (*S. literosa* subspecies; type locality, Wreck Bay, San Cristóbal)
> *S. punctata* Scudder 1893: 16 (*S. literosa* subspecies; type locality, Española)
> *S. hyalina* Scudder 1893: 16 (*S. literosa* subspecies; type locality, Genovesa); San Cristóbal
> Distribution. Endemic; Española, Floreana, Gardner near Española, Genovesa, San Cristóbal.
> Bionomics. Herbivore; arid zone; February–June. This is a smaller and more drab locust with a relatively shorter and wider hind femora, and spotted tegmina.

S. melanocera (Stål) 1860: 326 (*Acridium*); Butler 1877: 88 (*Acridium*); Howard 1889: 193; Scudder 1893: 11; McNeill 1901: 495; Hebard 1920: 323; Caudell 1931: 5; Dirsh 1969: 31; 1974: 113; Peck 1996: 1503.

> *S. minor* Snodgrass 1902: 421 (*S. melanocera* variety; type locality, Tagus Cove, Isabela)
> *S. pallida* Snodgrass 1902: 422 (*S. melanocera* variety; type locality, Santa Cruz
> *S. lineata* Snodgrass 1902: 423 (*S. melanocera* variety; type locality, Iguana Cove, Isabela)

S. immaculata Snodgrass 1902: 423 (*S. melanocera* variety; type locality, Santa Cruz)

S. intermedia Snodgrass 1902: 431 (type locality, Pinzón); Hebard 1920: 324

S. borealis Snodgrass 1902: 435 (*S. intermedia* variety; type locality, Marchena)

Distribution. Endemic; Galápagos: widespread (on all islands except Española), Baltra, Fernandina, Floreana, Gardner at Floreana, Genovesa, Isabela, Marchena, Pinzón, Pinta, Rábida, Santa Cruz, Santiago, Santa Fé, San Cristóbal

Bionomics. Herbivore; arid to humid forest zone; January–December. This is a large locust, with rather uniformly coloured tegmina, and is common at lights. Adults are most common in the rainy season from March to May.

Suborder Ensifera
Family Tettigoniidae
The Long-horned Grasshoppers

Key to Galápagos species of Tettigoniidae

1a Pronotum about as long as wide, surface rugose or coarsely granulate; with two transverse grooves; wings reduced; body stout; (Pseudophyllinae) . . . . *Nesoecia cooksoni*

1b Pronotum longer than wide, surface smooth or finely granulate, without two transverse grooves; wings either reduced or fully developed; body more elongate. 2

2a Dorsal surface of first tarsal segment smoothly rounded; hind wings fully developed, hind wings longer than front wings; length including wings about 25 cm; body light green (Phaneropterinae) . . . . *Anaulocomera darwinii*

2b Dorsal surface of first tarsal segment grooved laterally; other characters not exactly as above 3

3a Anterior portion of head vertex a rounded tubercle, not conspicuously projecting forward beyond antennal segment; body smaller, 25–30 mm long; wings not fully formed in adult (Conocephalinae) . . . . *Conocephalus exitosus*

3b Anterior protion of head vertex projecting conspicuously forward beyond basal antennal segment; body larger, 45–60 mm long including fully formed wings (Copiphorinae) 4

4a Head vertex smoothly rounded between eyes, with dark bar crossing just below vertex; body green or grey-brown . . *Neoconocephalus triops*

4b Head vertex sharply projecting and pointed; without lateral dark bar
below vertex; body green . . . . *Copiphora brevicauda*

Subfamily Copiphorinae
The Cone-headed Grasshoppers
Genus *Neoconocephalus* Karny 1907

N. triops (Linnaeus) 1758: 430 (*Gryllus-Tettigonia*); Hebard 1920: 337; Peck
1996: 1505.

> *Conocephalus insulanus* Scudder 1893: 21 (not Redtenbacher 1891);
> McNeill 1901: 501.
> *Conocephalus insularum* Karny 1907: 38 (replacement name for Scudder's
> preoccupied name).
> Distribution. Indigenous; North America to northern South America;
> Española, Fernandina, Isabela (Caleta Tagus), Pinta, San Cristóbal,
> Santiago (La Tragica, 360 m), Santa Cruz (Academy Bay).
> Bionomics. Herbivore; arid to humid forest zone; February–June. Females
> are green and males are brown-grey.

Genus *Copiphora* Audinet-Serville 1831

C. brevicauda Karny 1912: 12; Peck 1996: 1505.

> Distribution. Adventive; Neotropical. The species is not reported for Ecua-
> dor by Hebard (1924), but the type locality is Cachabi, Esmeraldas
> Province, Ecuador; Isabela, San Cristóbal, Santa Cruz.
> Bionomics. Herbivore; transition, agriculture and evergreen shrub zones;
> February–June. The species has been intercepted on bananas at U.S.
> ports (specimens in USNM), so it could have travelled to the Galápagos
> on supply boats, after flying to the boat's lights at night.

Subfamily Pseudophyllinae
The True Katydids
Tribe Cocconotini
Genus *Nesoecia* Scudder 1893

The Galápagos species has sometimes been placed in the genus *Liparoscelis*
Stål 1873, but this genus name is now restricted to *L. pallidispina* Stål 1873,
from Mexico (Bier 1960). *Nesoecia* also has species in Mexico, the Yucatan, and
coastal Brazil (Bier 1960). Colonization of the Galápagos from Pacific coastal
Mexico or Central America is probable. This genus is flightless everywhere and
must have dispersed by rafting.

N. cooksonii (Butler) 1877: 87 (*Agroecia*); Scudder 1893: 21; McNeill 1901: 497; Hebard 1920: 332 (all *Nesoecia*); Caudell 1931: 4; Parkin et al. 1972: 102; Peck 1996: 1504 (as *Liparoscelis*).

> *Nesoecia cooksoni ensifer* McNeill 1901: 498 (synon. *fide* Hebard, 1920; type locality, Española)
>
> *Nesoecia pallidus* McNeill 1901: 500 (synon. *fide* Hebard, 1920; type locality, Santa Fé)
>
> *Nesoecia paludicola* McNeill 1901: 499 (synon. *fide* Hebard, 1920; type locality, Isabela)
>
> Distribution. Endemic; Campéon, Española, Floreana, Genovesa, Isabela (Caleta Tortuga and Caleta Tagus), Santa Cruz, Santiago.
>
> Bionomics. Herbivore; littoral (in mangroves) to humid forest and pampa zone; January–November; often found in hiding under leaves of *Tournefortia* shrubs, and in woody debris on forest floors, and in hollows and cracks in standing dead or live trees. The wide geographic and ecological distribution suggests that there may be more than one species, and that the names of McNeill may be valid.

Subfamily Conocephalinae
The Meadow Grasshoppers
Genus *Conocephalus* Thunberg 1815

C. exitiosus (McNeill) 1901: 501 (*Xiphidium*); Hebard 1920: 335; Peck 1996: 1505.

> Distribution. Endemic; Santa Cruz, Santiago, San Cristóbal (3 in MCZ, El Junco, 15–16 IV. 1970, R. Silberglied)
>
> Bionomics. Herbivore; arid zone; April–May; with functional flight wings.

C. new species A

> Distribution. Endemic; Santa Cruz
>
> Bionomics. Herbivore; humid forest and pampa zone; March and May; with reduced flight wings.

C. new species B

> Distribution. Endemic; San Cristóbal
>
> Bionomics. Herbivore. *Miconia* and pampa zones; with reduced flight wings.

Subfamily Phaneropterinae
The Katydids
Genus *Anaulocomera* Stål 1873

A. darwinii Scudder 1893: 19; McNeill 1901: 496; Hebard 1920: 332; Peck 1996: 1505.

A. cornucervi Howard 1889: 192 (not Brunner)

Distribution. Endemic; Isabela, San Cristóbal, Santa Cruz, Santiago (Los Guayabillos, 230 m)

Bionomics. Herbivore; arid to humid forest zone; January–June; found in and on shrubs.

Family Gryllidae
The Crickets

Key to the Galápagos species of Gryllidae

1a Hind tibiae without long spines (but with apical spurs); body covered with scales; hind femora stout (Mogoplistinae). . . . *Cycloptilum* species.

1b Hind tibiae with long spines; body with setae, not scales; hind femora only moderately enlarged 2

2a Second tarsal segment larger, somewhat expanded laterally, flattened dorsoventrally; hind tibiae without small teeth between longer spines; ovipositor compressed, upcurved (Trigonidiinae) *Jarmilaxipha ecuadorica*

2b Second tarsal segment small, flattened laterally; hind tibiae with or without small teeth between spines; ovipositor not compressed, curved upward (Gryllinae). 3

3a Body larger, over 14 mm long; body dark brown or black; tegmina and wings shorter than, as long as, or longer than abdomen; hind tibiae with 6 or more pairs of spines before apex . . . . *Gryllus*

3b Body smaller, less than 14 mm long; body lighter in colour, but with dark areas; tegmina shorter than 1/2 length of abdomen, hind wings absent; hind tibiae with 5 or fewer pairs of spines before apex 4

4a Hind tibiae with 6 outer dorsal and 5 inner dorsal spines, and apex with 2 outer and 3 inner spines; last segment of maxillary palps only slightly longer than preceding segment . . . . *Gryllodes sigillatus*

4b Hind tibiae with 5 or fewer pairs of long dorsal spines; last segment of maxillary palps longer than preceding segment (Nemobiinae) 5

5a   Hind tibiae with 4 inner and 4 outer dorsal spines and 3 outer and 4 in-
     ner distal spines (spurs); *Pteronemobius* 8
5b   Hind tibiae with other spine numbers 6
6a   Eyes faceted, large; body colour dark; coastal inhabitant;
     Hygronemobius. 7
6b   Eyes unfaceted, smaller; body colour light tan; in caves or a transition
     to pampa zone rock talus inhabitant; hind tibiae with 5 outer and 5 in-
     ner dorsal spines (1 and 5 may be small or absent) and 3 outer and 3 in-
     ner dorsal spines . . . . *Anurogryllus typhlops* and species 2
7a   Larger, adult length 12–13 mm; dark brown to black on dorsum, pale
     on venter, legs speckled white and black; male forewings about as long
     as pronotum, dark brown; on all islands in coastal rocky zones
     . . . . *Hygronemobius speculi*
7b   Smaller, adult length 7–8 mm; mostly dark on dorsum, strongly speck-
     led; all legs strongly banded; male forewings about twice as long as
     pronotum, black with pale veins; mudflats under mangroves
     . . . . *Hygronemobius daphne*
8a   File teeth of male wing with abrupt transition from narrow to wide
     spacing in middle of the file . . . . . . *Pteronemobius santacruzensis*
8b   File teeth of male wing gradually changing in spacing along the length
     of the stridulatory file . . . . *Pteronemobius cristobalensis*

Subfamily Trigonidiinae
The Bush Crickets
Genus *Jarmilaxipha* Otte and Peck 1998

J. ecuadorica Otte and Peck 1998 *b*: 237; Peck 1996: 1506 (*Anaxipha peruviana*
Saussure?)

   Distribution. Adventive; probably recently accidentally introduced from
      mainland Ecuador; Floreana, Isabela, San Cristóbal, Santa Cruz, Santi-
      ago. Nine species of *Anaxipha*, a similar genus, are reported from main-
      land Ecuador (Hebard 1924), but this is distinctly different from them.
   Bionomics. Herbivore; maybe scavenger?; arid to pampa zone; February–
      June, and September. The species was probably carried to the
      Galápagos on living plants. It frequently comes to lights, so could also
      have been attracted to ship lights, and then carried to and throughout the
      Archipelago. The earliest records are from 1970 on Floreana
      (1.III.1970, Blackbeach, W.E. Ferguson, at light on ship, in USNM) and
      from Santa Cruz (43 in MCZ, 4.III.70, CDRS, R. Silberglied).

Subfamily Mogoplistinae
The Scaly Crickets
Genus *Cycloptilum* Scudder 1868

C. erraticum Scudder 1893: 23; McNeill 1901: 505; *Cryptoptilum erraticum* (Scudder); Hebard 1920: 338; Chopard 1968: 228 (*Ornebius*); Peck 1996: 1506.

C. lepismoide McNeill 1901: 505; Hebard 1920: 340 (as *Cryptoptilum*); Chopard 1968: 228 (*Ornebius*).

> Distribution of both above. Endemic; Fernandina, Floreana, Gardner, Española, Isabela, Española, Marchena, Pinta, Pinzón, Rábida, San Cristóbal, Santa Cruz, Santiago, Genovesa.
> Bionomics of both above. Herbivore; scavenger?; littoral to pampa zone. Adult males retain forewings for stridulating, adult females have no wings. Both sexes are flightless.
> Notes. I have been unable to separate my collections (now deposited in CAS) into the two above named species. These crickets are more readily separated or characterized by song than by morphology. In the field the songs indicated that more than two species were present. This offers a research opportunity in this notoriously difficult group of crickets. Chopard (1968: 228) notes *E. erraticum from* "Maria Madre Island" and this may refer to Isla Floreana.

Subfamily Nemobiinae
The Ground Crickets
Genus *Anurogryllus* Saussure 1877

Anurogryllus typhlops Otte and Peck 1998a: 227; Peck 1996: 1506.

> Distribution. Endemic; Isabela.
> Bionomics. Scavenger?; under moss mats and deep in rock piles on Sierra Negra, 750–1000 m (pampa); eyeless; also in Cueva Sucre at Santo Tomás; and at 200 m (3 km W Caleta Iguana), Cerro Azul (in pan trap). This and the following eyeless species have no eyed surface-dwelling relatives in the Galápagos. The original eyed colonizing continental ancestor probably independently occupied subsurface habitats on the two different islands as an escape from an episode of extreme dryness, which caused the extinction of the surface ancestors.

Anurogryllus species 2, Otte and Peck 1998a: 228; Peck 1996: 1506.

> Distribution. Endemic; Santa Cruz.
> Bionomics. Scavenger?; in caves in agricultural zone; known only from one nymph; eyeless.

Genus *Hygronemobius* Hebard 1913

H. daphne Otte and Peck 1998*b*: 231

> Distribution. Adventive; natural distribution probably mainland Ecuador; Santa Cruz.

> Bionomics. Scavenger; littoral zone; on mud exposed at low tide, under mangroves near Park Headquarters; January–April; first discovered in 1996; taken by sweeping over tidal mud or in pan traps; not found in mangroves east of Villamil on Isabela.

H. speculi (McNeill) 1901: 503 (*Nemobius*); Hebard 1920: 341; Otte and Peck 1998*b*: 233.

> Distribution. Endemic; Darwin, Española, Fernandina, Floreana, Isabela, San Cristóbal, Santa Fé; to be expected on every island.

> Bionomics. Scavenger?; littoral and arid zone; January–July; caught by spraying cracks in sea shore rock cliffs with short lived pyrethrum insecticide. Males can be heard stridulating in cracks in sea shore cliffs. Both sexes are flightless, as are nemobines in general and all such indigenous and endemic island crickets probably dispersed by rafting.

Genus *Pteronemobius* Audinet-Serville 1839

Pteronemobius cristobalensis Otte and Peck 1998*b*: 235; Peck 1996: 1506 (*Nemobius*).

> Distribution. Endemic; San Cristóbal.

> Bionomics. Scavenger?; *Miconia* and pampa zones only, 550–675 m elevation, around El Junco region; February–March.

Pteronemobius santacruzensis Otte and Peck 1998*b*: 235; Peck 1996: 1506 (*Nemobius*).

> Distribution. Endemic; Santa Cruz.

> Bionomics. Scavenger; pampa zone only, 600–800 m. February–June; often caught in pit traps. Wings are reduced and function only in stridulating in the males. The species are very similar and do not differ in genitalia, but do differ in calling songs. This genus probably colonized in a flightless condition and both species descended from a common ancestor. The ancestor may have colonized in the arid zone, but the daughter species now occur only in the upper pampa zone of the two separate isalands.

Subfamily Gryllinae
The Field Crickets
Genus *Gryllodes* Saussure 1874

G. sigillatus (Walker); Peck 1996: 1507. The decorated cricket.
   Distribution. Adventive; circumtropical; Genovesa, Marchena, Santa Cruz
      (CDRS), San Cristóbal.
   Bionomics. Herbivore, maybe a scavenger; arid zone, at night on paths;
      males are frequently heard singing their distinctive calling song at night
      in coastal villages; March–May.

Genus *Gryllus* Linnaeus 1758

Identification of Galápagos *Gryllus* species is made relatively simple by the
fact that each island contains only one or two species. However, there is the possibility that species will have wider distributions than we now know or that other
species will be discovered.

Key to the *Gryllus* of the Galápagos (from Otte and Peck 1997)

1a   From Isla Darwin . . . . *G. darwini*
1b   From Isla Genovesa . . . . *G. genovesa*
1c   From Isla Marchena . . . . *G. marchena*
1d   From other islands 2
2a   From Isla Pinta. 3
2b   Not from Isla Pinta 4
3a   Legs 1 and 2 light brown, without dark marking; hind femora orange
       brown; back of head without bands . . . . *G. abingdoni*
3b   Legs 1 and 2 pale, with dark streaks and spots; hind femora pale with
       strongly defined oblique dark stripes . . . . *G. pinta*
4a   Back of head striped; lower front corner of lateral lobe of pronotum
       with a pale area; hind legs pale with dark brown or black stripes;
       hindwings extending beyond the abdomen; San Cristóbal, Española,
       Fernandina, Floreana, Isabela, Santa Cruz, Santiago . . . . . *G. abditus*
4b   Back of head black. Lower front corner of lateral lobe of pronotum not
       pale. Hind legs more or less uniform in colour (orange to dark reddish
       brown). 5
5a   Ovipositor rather long, body length larger; Santa Cruz, Santa Fé
       . . . . *G. galapageius*
5b   Ovipositor short, body small; Fernandina, Isabella . . . . . . *G. isabela*

G. abditus Otte and Peck 1997: 165.
> Distribution. Endemic; Española, Floreana, Fernandina, Isabela, San Cristóbal, Santa Cruz, Santiago. The wings are both macropterous and micropterous (reduced and not functional for flight).
> Bionomics: Herbivore; arid to pampa and agriculture zones.

G. abingdoni Otte and Peck 1997: 171.
> Distribution. Endemic; Pinta.
> Bionomics. Herbivore; arid zone; March; flightless; shares island with *G. pinta*.

G. darwini Otte and Peck 1997: 173.
> Distribution. Endemic; Darwin.
> Bionomics. Herbivore; arid zone; May; flightless.

G. galapageius Scudder 1893: 22; McNeill 1901: 502 (synonomized with *G. assimilis* by Rehn and Hebard 1915: 296); Hebard 1920: 342; Caudell 1931: 5 (all as *Gryllus assimilis* (Fabr.)); Peck 1996: 1506; Otte and Peck 1997: 165.
> Distribution. Endemic; Santa Cruz, Santa Fé; the type is from "Albemarle" (Isabela) but we know of no other specimens from Isabela, and suspect mislabeling.
> Bionomics. Herbivore; arid to humid forest zone; January, May, June, October. Wings are seemingly always fully developed and can be used for flight.

G. genovesa Otte and Peck 1997: 166.
> Distribution. Endemic; Genovesa.
> Bionomics. Herbivore; arid zone; March; flightless.

G. isabela Otte and Peck 1997: 166.
> Distribution. Endemic; Isabela.
> Bionomics. Herbivore; arid to pampa zone; February–September; flightless.

G. marchena Otte and Peck 1997: 166.
> Distribution. Endemic; Marchena.
> Bionomics. Herbivore; arid zone; March; flightless.

G. pinta Otte and Peck 1997: 171.
> Distribution. Endemic; Pinta.
> Bionomics. Herbivore; transition and humid forest; March; flightless; shares island with *G. abingdoni* and the species may separate their distributions by elevation (vegetational zones).

Acknowledgements

I thank Dan Otte (Academy of Natural Sciences, Philadelphia, PA) for information on Galápagos Orthoptera and pleasant and informative companionship in the field in the Galápagos and elsewhere.

References

Abs, M., Curio, E., Kramer, P., and Niethammer, J. 1965. Ernährungsweise der Eulen auf Galápagos Ergebnisse der Deutschen Galápagos-Expedition 1962/63. IX. J. Ornithol. **106**: 49–57.

Bier, M. 1960. Das Tierreich, Liefrung 74. Orthoptera, Tettigoniidae (Pseudophyllinae II). W. de Gruyter & Co., Berlin. pp 1–396.

Borror, D.J., Triplehorn, C.A., and Johnson, N.F. 1989. An introduction to the study of insects. 6th ed. Saunders College Publ., Philadelphia, PA. 875 pp.

Butler, A.G. 1877. Account of the zoological collection made during the visit of H.M.S. "Peterel" to the Galápagos Islands. X. Lepidoptera, Orthoptera, and Hemiptera. Proc. Zool. Soc. London 1877: 86–91.

Caudell, A.N. 1932. Insects of the Order Orthoptera of the Pinchot Expedition of 1929. Proc. U. S. Natl. Mus. 2921: 80: (art. 21): 1–7.

Chapman, R.F., Espelie, K.E., and Peck, S.B. 2000. Cuticular hydrocarbons of grasshoppers from the Galapagos Islands, Ecuador. Biochem. Syst. Ecol. **28**: 579–588.

Chopard, L. 1968. Orthopterorum Catalogus, pars 12. Uitgeverij Dr. W. Junk n.v. 's-Gravenhage, Netherlands.

Dirsh, V.M. 1969. Acridoidea of the Galápagos Islands (Orthoptera). Bull. Br. Mus. (Nat. Hist.) Entomol. **23**: 27–51, 75 text Figures, 1 map, 1 graph, pls. 1–7.

Dirsh, V.M. 1974. Genus *Schistocerca* (Acridomorpha, Insecta). Ser. Entomol. **10**: 1–238.

Grant, P.R. 1986. Ecology and evolution of Darwin's finches. Princeton University Press, Princeton, NJ. 458 pp.

Hebard, M. 1920. Expedition of the California Academy of Sciences to the Galápagos Islands, 1905–1906. XVII. Dermaptera and Orthoptera. Proc. Calif. Acad. Sci. ser. 4, 2: 311–346, pl. 18, t. Figs. 1–11.

Hebard, M. 1924. Studies in the Dermaptera and Orthoptera of Ecuador. Proc. Acad. Nat. Sci. Phila. **76**: 109–248.

Hebard, M. 1934. The Norwegian Zoological Expedition to the Galápagos Islands 1925, conducted by Alf Wollebaek. X. Orthoptera of the Galápagos Islands. Nyt Magazin Naturvidenskaberne, vol. 74. pp. 279–280.

Howard, L. 1889. Scientific results of explorations by the U.S. Fish Commission Steamer Albatross. no. V. Annotated catalogue of the insects collected in 1987–88. Proc. U.S. Natl. Mus. **12** (no. 771): 185–216.

McNeill, J. 1901. Papers from the Hopkins Stanford Galápagos Expedition, 1898–1899. Entomological Results (4): Orthoptera. Proc. Wash. Acad. Sci. **3**: 487–506.

Otte, D. 1989. Speciation in Hawaiian crickets. *In* Speciation and its consequences. *Edited by* D. Otte and J.A. Endler. Sinauer Assoc., Sunderland, MA. pp. 482–526.

Otte, D., and Peck, S.B. 1997. New species of *Gryllus* (Orthoptera: Grylloidea; Gryllidae) from the Galápagos Islands. J. Orthoptera Res. **6**: 161–174.

Otte, D., and Peck, S.B. 1998*a*. A new blind *Anurogryllus* from the Galápagos Islands, Ecuador (Orthoptera: Gryllidae: Brachytrupinae). J. Orthoptera Res. **7**: 227–229.

Otte, D., and Peck, S.B. 1998*b*. Crickets from the Galápagos Islands, Ecuador (Orthoptera, Gryllidae, Nemobiinae). J. Orthoptera Res. **7**: 231–240.

Parkin, P., Parkin D.T., Ewing, A.W. and Ford, H.A. 1972. A report on the arthropods collected by the Edinburgh University Galápagos Islands expedition, 1968. The Pan-Pacific Entomol. **48**: 100-107.

Peck, S.B. 1996. Diversity and distribution of the Orthopteroid insects of the Galápagos Islands, Ecuador. Can. J. Zool. **74**: 1497–1510.

Rehn, J.A.G., and Hebard, M. 1915. The genus *Gryllus* (Orthoptera) as found in America. Proc. Acad. Nat. Sci. Phila. **67**: 292–322.

Scudder, S.H. 1893. Reports on the dredging operations off the west coast of Central America to the Galápagos, to the west coast of Mexico, and in the Gulf of California, in charge of Alexander Agassis, carried on by the U.S. Fish Commission Steamer "Albatross" during 1891, Lieut. Commander Z.L. Tanner, U.S.N., Commanding. VII. The Orthoptera of the Galápagos Islands. Bull. Mus. Comp. Zool. (Harvard) **25**: 1–25, pls. I–III.

Sjöstedt, Y. 1932. Orthopteren in Naturhistorischen Reichmuseum zu Stockholm. 2. Acrididae. Ark. Zool., ser. A, **24**: 1–89, 20 pls.

Snodgrass, R.E. 1902. Papers from the Hopkins Stanford Galápagos Expedition, 1898–1899. VIII. Entomological Results (7): *Schistocerca, Sphingonotus,* and *Halmenus.* Proc. Wash. Acad. Sci. **4**: 411–454, pls. XXVI–XXVII.

Stål, C. 1860. Orthoptera species novas descripsit. *In* Kongliga Svenska Fregatten Eugenies Resa Omkring Jorden, under befäl af C.C. Virgen, aren 1851–3. Zoologi 1, Insecta: 299–350, 1 pl.

Walker, F. 1870. Catalogue of the specimens of Dermaptera Saltatoria in the collection of the British Museum 4: 605–809.

Chapter 11
Order Mantodea

The Praying Mantises

1 family, 1 genus (endemic), 1 species (endemic)
Minimally 1 natural colonization

There are eight families of Mantids worldwide. Most families, genera, and species are tropical. All are predators on other insects. While the fauna of tropical America is very rich, there is only one species on the Galápagos. They are most often seen at night at lights or by beating vegetation. Many mantid females are flightless, as is the Galápagos species, but I assume this condition did not arise on the Galápagos. Dispersal was probably by rafting of a female or juveniles, but oöthecae (egg masses) could also occur on rafting debris.

Family Mantidae
Genus *Galapagia* Scudder 1893 (endemic genus)

G. solitaria Scudder 1893: 8; Hebard 1920: 317.
   Vates sp Butler; 1877: 88 (2 nymphs from Floreana); Scudder 1893: 8.
   Distribution. Endemic; Baltra, Española, Floreana, Isabela, San Cristóbal, Santa Cruz, Santiago, Seymour.
   Bionomics. Arid to humid forest zones; January–March, June, September–October. The adult males have functional wings, but adult females have much reduced wings. Collected by beating vegetation, in malaise traps, or at lights.
   Phylogeny. The genus is supposed to be closely related to *Musonia* Stål (= *Thespis* Saussare), and between it and *Brunneria* (Scudder 1893: 8). This question needs to be reexamined.

References

Butler, A.G. 1877. Account of the zoological collection made during the visit of H.M.S. "Peterel" to the Galápagos Islands. X. Lepidoptera, Orthoptera, Hemiptera. Proc. Zool. Soc. London 1877: 86–91.

Hebard, M. 1920. Expedition of the California Academy of Sciences to the Galápagos Islands, 1905–1906. XVII. Dermatera and Orthoptera. Proc. Calif. Acad. Sci. ser. 4, 2: 311–346, pl. 18, t.Figs. 1–11.

Scudder, S.H. 1893. Reports on the dredging operations off the west coast of Central America to the Galápagos, to the west coast of Mexico, and in the Gulf of

California, in charge of Alexander Agassiz, carried on by the U.S. Fish Commission Steamer "Albatross" during 1891, Lieut. Commander Z.L. Tanner, U.S.N., Commanding. VII. The Orthoptera of the Galápagos Islands. Bull. Mus. Comp. Zool. Harvard **25**: 1–25, pls. I–III.

Chapter 12
Order Blattodea

The Cockroaches

4 families, 12 genera, 18 species (5 endemic, 2 indigenous, 11 adventive)
Minimally 3 (or 4) natural colonizations

Cockroaches are archaic, hardy, and successful insects and are most diverse in tropical countries. They are usually general feeders and many do well in the presence of humans. Seemingly only three or four species naturally colonized the Galápagos on their own, by rafting. Considering the rich Neotropical fauna, this seems a surprisingly low number. Another 11 species were probably brought by humans, some perhaps as early as the first European landing in 1535 of the Bishop of Panama, Tomas de Berlanger. The adventive species *Periplaneta australasiae* has become very abundant in natural habitats in arid and transition zone forests, where it can be seen commonly at night. Its impact on the fauna and flora is unknown. *Symploce pallens* has females with reduced wings, is very abundant in littoral habitats, and was probably adventive from tropical America where it is a widespread indigenous species. Most of the other adventive species are ultimately from the Old World tropics.

Cockroaches in the coastal and upland villages may now be of serious potential medical importance. They may be able to transmit communicable diseases from open sewage and human waste onto food or into water. The best cockroach control measures are (1) cleanliness, and (2) by making traps with baits hanging into smooth-walled glass jars with some cooking oil above a layer of water in which they drown. The use of chemical poisons for cockroach control is not advised because they may affect too many neutral or beneficial insects and other animals such as geckoes. The best cockroach predators are the giant Galápagos centipede and giant huntsman house spider, and they should not be killed for this reason. Evaniid ensign wasps are parasitoids of *Periplaneta* cockroach eggs, and these have recently appeared in the Galápagos.

Adult cockroaches usually have fully formed tegmina and hindwings, as well as well-formed and sclerotized genitalia. Females are usually larger and stouter than males of the same species. Flightless adults have the tegmina and wings reduced in size to varying degrees, always exposing all or part of the abdomen. When tegmina and wings are reduced to mere flaps these are often flexibly joined to the thorax. When tegmina and wings are completely absent, the recognition of the adult stage must depend upon the presence of fully developed genitalia. Reduction of flight organs may vary between the sexes and wings are then usually reduced or lost in the female, as in *Symploce pallens*. Immature cockroaches (not covered in the following keys) resemble the adults except for their smaller size, lack of functional genitalia, and the absence of fully formed fore and hind wings. Nymphs have poorly sclerotized and undeveloped genital phallomeres. Early instars of both sexes have styles on the subgenital plate. Immature males retain the styles in all instars and also possess them as adults. Older female instars lose the styles and lack them in the adult stage. All nymphal stages can be sexed by differences in the subgenital plate. Wing pads may be noted in older nymphs as simple, non-flexible, lateral lobe-like extensions of the posterior margins of the wing-bearing thoracic segments. A detailed review of Galápagos cockroaches, with illustrations, is that of Peck and Roth (1992).

Key to adult cockroaches of the Galápagos Islands

Additional cockroach species may remain to be discovered or may yet be introduced by commerce from Central or South America. Keys in Helfer (1963), Borror et al. (1989), Fisk and Wolda (1979) and Fisk (1982) may be helpful in identifying further adventive species

Instead of lengthening this chapter by providing illustrations which have appeared elsewhere, we make reference with an asterisk symbol (*) to habitus illustrations in Hebard (1917) and with a double asterisk (**) to habitus photographs in Roth and Willis (1960). Although colour markings often are distinctive, they are variable, and should be used with caution. Other illustrations are in Peck and Roth (1992).

1a Body length less than 7 mm. 2
1b Body length 7 mm or more 4
2a Pronotum setose; tegmina (fore-wings) with proximal region sclerotized, opaque, setose; distal portion unsclerotized, hyaline. (Pl. VIII, Figs. 8–11*); Polyphagidae.
. . . . *Holocompsa* species and *Holocompsa nitidula*
2b Pronotum not setose and tegmina not as above; hind wing, if fully developed, with a large apical triangle which is reflexed over the rest of the wing when at rest; Blattellidae. 3
3a Pronotal disk very dark, lateral borders hyaline, white; tegmina and wings fully developed, the former very dark with white anterior margin; male supraanal plate with medial setal specialization.
. . . . *Anaplecta lateralis*

3b Pronotal disk not as above; tegmina and wings either fully developed or reduced; if the former, the tegmina are pale, hyaline; male abdominal tergum 7 specialized (*Chorisoneura*) (see also couplet 10) 13

4a Male subgenital plate with a pair of similar, elongate cylindrical styles situated laterodistally; female subgenital plate with mesodistal portion valvular; Blattidae 5

4b Male and female subgenital plates not as above 7

5a Humeral area of tegmina (region anterior to the subcosta) yellow; pronotum reddish brown, tinged with orange, with 2 large sharply defined, often confluent black blotches (male and female) Pl. 20**); hind margin of supraanal plate subtrapezoidal (male), or apically excavated (female) . . . . *Periplaneta australasiae*

5b Humeral region of tegmina not yellow; male supraanal plates not as above 6

6a Supraanal plate largely unsclerotized, membranous, hind margin with a V-shaped notch or emargination (male and female); pronotal pattern variable (Pl. 19**) . . . . *Periplaneta americana*

6b Male supraanal plate with hind margin weakly convex, corners rounded; female supraanal plate apically indented; pronotum dark reddish brown (variable) (Pl. 21**) . . . . *Periplaneta brunnea*

7a Cerci long, slender, extending well beyond supraanal plate; legs long, slender; Blattellidae 8

7b Cerci short, not projecting much beyond hind margin of subgenital plate; legs short and stocky; Blaberidae 4

8a Anteroventral margin of front femur armed with a row of stout spines which decrease in length distally, terminating in 3 larger spines that increase in length distally. 9

8b Front femur not as above 10

9a Pronotum with a pair of dark longitudinal bands (male and female) (Pl. 5, A, B**); tegmina and wings fully developed; male abdominal tergum 7 and 8 with specialized structure (see Roth 1985, Fig. 4 B) . . . . *Blattella germanica*

9b Pronotum without longitudinal bands; male with fully developed tegmina and wings (Roth 1984, Fig. 15A), and only seventh abdominal tergum specialized (Roth 1984, Fig. 17B); tegmina and wings of female reduced (see Roth 1984, Fig. 15B) . . . . *Symploce pallens*

10a Anteroventral margin of front femur armed with 2–3 large medioproximal spines, succeeded by a row of small, uniform, piliform spinules, terminating in 2–3 large spines; both seventh and eighth male abdominal terga specialized; *Ischnoptera* 11

10b Anteroventral margin of front femur without large proximal spines, terminating in 1 large spine, with uniformly short spinules only; tegmina and hind wings reduced, short, not covering abdomen; male seventh abdominal tergite with specialized gland area; *Chorisoneura* 13

11a Eyes absent; tegmina reduced to small lateral pads, hind wings absent (see Roth 1988, Fig. 4A) (male); (female unknown)
. . . . *Ischnoptera peckorum*

11b Eyes present; tegmina larger, width normal, but reaching only to about abdominal tergum 1 or 2, hind wings vestigial 12

12a Hind margin of supraanal plate weakly indented (male) or trigonal, apex rounded (female) . . . . *Ischnoptera santacruzensis*

12b Hind margin of supraanal plate deeply and widely excavated (male), or trigonal, apex rounded (female) . . . . *Ischnoptera snodgrassii*

13a Specialization on male abdominal tergum 7 a semicircular fossa containing numerous setae; tegmina and wings fully developed or somewhat reduced; if the latter, then not quite reaching end of abdomen (wings shorter) (male and female) . . . . *Chorisoneura carpenteri*

13b Specialization on male abdominal tergum 7 with a triangular raised mound or ridge which divides a setose fossa whose anterior margin is heart-shaped; tegmina reduced, reaching to abdominal tergum 5 or 6, hind wings vestigial (male and female) . . . *Chorisoneura cristobalensis*

14a Anteroventral margin of front femur with 3–5 stout spines in middle followed by a row of short setae, or piliform setae, terminating in a stout spine 15

14b Anteroventral margin of front femur with a row of fine setae only, heavy proximal spines absent 16

15a Terminal spine on the anteroventral margin of front femur about the same size as the stout medial spines; ventral margins of mid and hind femora without stout spines; pronotal disk with a black shield-shaped macula; humeral vein of tegmina black; body length, 45–50 mm. (see Roth 1969, Fig. 154). . . . . *Blaberus parabolicus*

15b Terminal spine on the anteroventral margin of front femur much longer than the stout medial spines; ventral margins of mid and hind femora with a few widely spaced spines; pronotal disk with a large brownish red macula whose delimiting margin is undulate; body length about 15 mm. (Pl. VI, Fig. 9*). . . . . *Phoetalia pallida*

16a Anteroventral margin of front femur with a single, stout, terminal spine; proximal part of tegmina densely punctate; pronotum blackish or reddish brown with pale anterolateral borders (Pl. VIII, Fig. 1*; Pl. 24B**) . . . . *Pycnoscelus surinamensis*

16b Anteroventral margin of front femur without a stout spine. Pronotum
   not as above, base colour ash-grey 17
17a Pronotum subparabolic, with a dark black or brownish stripe along the lateral borders, and area between them a symmetrical lighter brown pattern;
   tegmina and wings reaching to about end of abdomen; body length about
   22–30 mm (male and female) (Pl. 14**) . . . . *Nauphoeta cinerea*
17b Pronotum subelliptical, with a few small symmetrical spots; tegmina
   and wings extend beyond end of abdomen; body length 38–53 mm
   (male and female) (Pl. 13**) . . . . *Rhyparobia maderae*

Family Blattidae
Genus *Periplaneta* Burmeister 1838

P. americanna (Linnaeus) 1758: 424; Butler 1877: 87; Howard 1889: 193;
Scudder 1893: 6; McNeill 1901: 494; Hebard 1934: 279; Peck and Roth 1992:
2205. The American cockroach.
   Distribution. Adventive; cosmopolitan; Floreana, Genovesa, Isabela, San
      Cristóbal, Santa Cruz.
   Bionomics. Scavenger; frequently occurring around houses; collected
      February–May, July, and September–December. It is now known from
      the littoral zone to humid forest at 380 m, and in the agriculture zone
      and caves on Santa Cruz at Bellavista and El Chato.

P. australasiae (Fabricius) 1775: 271; Howard 1889: 194; Scudder 1893: 6; Hebard
1920: 316; 1934: 279; Parkin et al. 1972: 102; Peck and Roth 1992: 2205. The Aus-
tralian cockroach.
   Distribution. Adventive; cosmopolitan; Floreana, Isabela, San Cristóbal,
      Santa Cruz.
   Bionomics. Scavenger; frequently occurring around houses; littoral to hu-
      mid forest zones; abundant at night on trees in transition zone forest on
      Santa Cruz; collected January–November.

P. brunnea Burmeister 1838: 503; Hebard 1920: 316; Peck and Roth 1992: 2205.
The Southern brown cockroach.
   Distribution. Adventive; circumtropical; San Cristóbal.
   Bionomics. Scavenger; frequently occurring around houses; collected in
      January. There are no recent records of the species.

Family Polyphagidae
Genus *Holocompsa* Burmeister 1838

H. nitidula (Fabricius) 1871: 345; Peck and Roth 1992: 2205. The small hairy
cockroach.

Distribution. Indigenous?; tropical America; Isabela (near Villamil), Santa Cruz.

Bionomics. Scavenger; a house-dwelling species that may occur in disturbed habitats; collected March–May in litter below adventive *Ceiba* trees and at uv lights in coastal arid zone.

H. species; Peck and Roth 1992: 2205.

Distribution. Indigenous ?, Santa Cruz.

Bionomics. Scavenger; in nests of *Xylocopa* bees; arid zone.

Family Blattellidae
Genus *Anaplecta* Burmeister 1838

A. lateralis (Burmeister) 1838: 494; Peck and Roth 1992: 2206.

Distribution. Adventive; Panama to Brazil; Santa Cruz.

Bionomics. Scavenger; arid to pampa and agriculture zone; at lights and in traps in forest; March–May; first discovered in 1991 and very common in pastures and forests by 1996.

Genus *Blattella* Caudell 1903

B. germanica (Linnaeus) 1767: 668; Hebard 1920: 315; Peck and Roth 1992: 2205. The German cockroach.

Distribution. Adventive; cosmopolitan (originating in eastern Asia); Española(?), Santa Cruz.

Bionomics. Scavenger; a species found in and around human habitation; collected in May–June, November and December. There is no human habitation on Española, so the early Española record (Hebard 1920) may have come from the ship of the collectors. I have found the species to be common on boats catering to the tourist trade operating out of Santa Cruz, and in buildings in Puerto Ayora, Santa Cruz. There are no records of this species having established itself on any other island.

Genus *Chorisoneura* Brunner v.W. 1865

C. carpenteri Roth, in Peck and Roth 1992: 2210.

Distribution. Endemic; Floreana, Isabela, Santa Cruz, Santa Fé, Santiago.

Bionomics. Scavenger; arid to pampa zones; in pit and pan traps; January–June. There is considerable variation in length of tegmina and flight wings on and between islands.

C. cristobalensis Roth, in Peck and Roth 1992: 2212.

Distribution. Endemic; San Cristóbal.

Bionomics. Scavenger; in humid forest and *Miconia* litter and pampa zone from 500–700 m; February. The tegmina is reduced and the hind wings are vestigial. This species is probably a sister species to the above one.

Genus *Ischnoptera* Burmeister 1838

I. peckorum Roth 1988: 307; Peck and Roth 1992: 2210.
  Distribution. Endemic; Santa Cruz; transition zone cave.
  Bionomics. Scavenger; an eyeless and wingless species from Bellavista Cave.
    This species is more closely related to *I. snodgrassi* of Isabella than it is to
    I. santacruzensis. It has not been seen since the original 1985 discovery.

I. santacruzensis Roth, in Peck and Roth 1992: 2210.
  Distribution. Endemic; Santa Cruz.
  Bionomics. Scavenger; in leaf litter of humid forest, evergreen shrub, and
    pampa habitats, 600–770 m; January–June; wings are reduced to small
    wing pads.

I. snodgrassii (McNeill) 1901: 493 (*Temnopteryx*); Hebard 1920: 315
(*Anisopygia*); Peck and Roth 1992: 2006.
  Distribution. Endemic; Isabela (Cerro Azul, Sierra Negra, Volcán Alcedo,
    Volcan Wolf).
  Bionomics. Scavenger; in humid forest and pampa habitats; February–June
    and September. Found under leaves on very wet ground covered with
    abundant and luxurious vegetation of ferns, shrubs and vines; usually
    under large rocks. Wings reduced to wing pads, probably descended
    from winged colonist. The flightless condition has not prevented it from
    spreading to all volcanos on Isabela.

Genus *Symploce* Hebard 1916

S. pallens (Stephens) 1835: 46; Roth 1984: 51; Peck and Roth 1992: 2206. The
smooth cockroach.
  S. lita Hebard 1916: 354; 1920: 316.
  S. hospes Hebard 1934: 279;
  Distribution. Adventive; circumtropical (widespread in Tropical America),
    originally from Africa; Bartolomé, Campéon, Fernandina, Floreana,
    Genovesa, Isabela, Marchena, Pinta, San Cristóbal, Santa Cruz, Santa
    Fé, Santiago.
  Bionomics. Scavenger; a household pest species; January–November; from lit-
    toral zone to humid forest at 650 m; abundant in arid zone around CDRS.

Family Blaberidae
Genus *Blaberus* Audinet-Serville 1831

B. parabolicus Walker 1868: 8; Peck and Roth 1992: 2214.
  Distribution. Adventive; Neotropics, Colombia to Surinam, Brazil and Peru;
    San Cristóbal (Baquerizo Moreno), Santa Cruz (Pto. Ayora).
  Bionomics. Scavenger; the species may occur around houses; and at lights in
    town; collected in February–March.

Notes. This is a taxonomically difficult genus and several species may eventually be adventive (Roth 1969).

Genus *Nauphoeta* Burmeister 1838

N. cinerea (Olivier) 1789: 314; Scudder 1893: 6; Hebard 1920: 317; Peck and Roth 1992: 2214. The cinereous cockroach.
> *N. bivittata* Burmeister (not Brunner); Howard 1889: 194.
> *N. circumvagans* Brun; in Scudder 1893: 7 (not Burmeister).
> *N. laevigata* ? (Palisot); Howard 1889: 194.
> Distribution. Adventive; circumtropical, probably originated in Africa; Floreana, San Cristóbal.
> Bionomics. Scavenger; the species often occurs around houses; probably an urban record, there are no habitat data for the collection; April, October. There are no recent records of the species.

Genus *Phoetalia* Stål 1874

P. pallida (Brunner) 1865: 286; Hebard 1920: 316 (*Leurolestes*); Peck and Roth 1992: 2214. The pallid cockroach.
> Distribution. Adventive; circumtropical, indigenous to West Indies; San Cristóbal.
> Bionomics. Scavenger; the species occurs in and around houses, probably an urban record; January. There are no recent records of this species.

Genus *Pycnoscelus* Scudder 1862

P. surinamensis (Linnaeus) 1767: 687; Butler 1877: 87 (*Panchlora*); Howard 1889: 194 (*Leucophaea*); Scudder 1893: 7; McNeill 1901: 494; (both *Leucophaea*); Hebard 1920: 317; Parkin et al. 1972: 102; Peck and Roth 1992: 2214. The Surinam cockroach.
> Distribution. Adventive; circumtropical, probably originating in Indo-Malayan region; Campéon, Española, Fernandina, Floreana, Gardner at Floreana, Isabela, San Cristóbal, Santa Cruz, Santiago.
> Bionomics. Scavenger; a parthenogenic species, often living near human habitation. Because it buries itself in soil, it can be carried in earth around plant stock. It is able to invade natural communities. Collected from January to June and from October to December in all vegetation zones from the littoral to the pampa (600 m) in both disturbed and undisturbed habitats.

Genus *Rhyparobia* Krauss 1892

R. maderae (Fabricius) 1781: 341; Peck and Roth 1992: 2241. The Madeira cockroach.

Distribution. Adventive; Old World tropics, from West Africa and now widely distributed in the Caribbean area; Floreana (Bahia Las Cuevas) San Cristóbal (Baquerizo Moreno).

Bionomics. Scavenger; the species may occur around houses; it was taken at lights in town and in rotted opuntia cactus; February, April.

Notes. This species is in older literature under the genus name *Leucophaea* (Kevan 1980).

References

Borror, D.J., Triplehorn, C.A., and Johnson, N.F. 1989. An introduction to the study of insects. 6th ed. Saunders College Publishing, Philadelphia.

Butler, A.G. 1877. Account of the Zoological Collection made during the visit of H.M.S. "Peterel" to the Galápagos Islands. X. Lepidoptera, Orthoptera, Hemiptera. Proc. Zool. Soc. London 1877: 86–91.

Fisk, F.W. 1982. Keys to the Cockroaches of Central Panama Part II: Flightless Species Studies on Neotropical Fauna and Environment. 17 pp. 123–127.

Fisk, F.W., and Wolda, H. 1979. Keys to the cockroaches of central Panama. Part I. Flying species. Stud. Neotrop. Fauna Environ. **14**: 177–201.

Hebard, M. 1917. The Blattidae of North America north of the Mexican boundary. Mem. Amer. Ent. Soc., No. 2, 284 pp., 10 pls.

Hebard, M. 1920. Expedition of the California Academy of Sciences to the Galápagos Islands, 1905–1906. XVII. Demaptera and Orthoptera. Proc. Calif. Acad. Sci. ser. 4, **2**: 311–346, pl. 18.

Hebard, M. 1934. The Norweigian Zoological Expedition to the Galápagos Islands 1925, conducted by Alf Wollebaek. X. Orthoptera of the Galápagos Islands. Nyt Magazin Naturvidenskaberne **74**: 279–280. [Reprinted as Meddelelser fra det Zoologiske Museum, Oslo, nr. **41**: 279–280 (1934).]

Helfer, J. 1963. How to know the grasshoppers, cockroaches and their allies. Pictured Key Nature Series. W.C. Brown, Dubuque, IA.

Howard, L.O. 1889. Scientific results of explorations by the U.S. Fish Commission steamer Albatross. No. V. Annotated catalogue of the insects collected in 1887–88. Proc. U.S. Natl. Mus. **12** (no. 771): 185–216.

Kevan, D.K.McE. 1980. Names involving the Madeira and Surinam cockroaches. Entomol. Rec. J. Variation **92**: 77–82.

McNeill, J. 1901. Papers from the Hopkins Stanford Galápagos Expedition, 1898–1899. Entomological Results (4); Orthoptera. Proc. Wash. Acad. Sci. **3**: 487–506.

Parkin, P., Parkin, D.T., Ewing, A.W., and Ford, H.A. 1972. A Report on the Arthropods Collected by the Edinburgh University Galápagos Islands Expedition, 1968. The Pan-Pacific Entomologist **48**: 100–107.

Peck, S.B., and Roth, L.M. 1992. Cockroaches of the Galápagos Islands, Ecuador, with descriptions of three new species (Insecta: Blattodea). Can. J. Zool. **70**: 2202–2217.

Roth, L.M. 1969. The male genitalia of Blattodea. I. *Blaberus* spp. (Blaberidae: Blaberinae). Psyche J. Entomol. **76**: 217–250.

Roth, L.M. 1984. The genus *Symploce* Hebard. I. Species from the West Indies (Dictyoptera: Blattodeae, Blattellidae). Entomol. Scand. **15**: 25–63.

Roth, L.M. 1985. A Taxonomic revision of the genus *Blattella* Caudell (Dictyoptera, Blatteria: Blattellidae) Ent. Scand. Supplement No. 22, pp. 221.

Roth, L.M. 1988. Some cavernicolous and epigean cockroaches with six new species, and a discussion of the Nocticolidae (Dictyoptera: Blattodea). Rev. Suisse Zool. **95**: 297–321.

Roth, L.M., and Willis, E.R. 1960. The Biotic Associations of Cockroaches, Smith Inc. 470 pp. publication 4422.

Scudder, S.H. 1893. Reports on the dredging operations off the west coast of Central America to the Galápagos, to the west coast of Mexico, and in the Gulf of California, in charge of Alexander Agassiz, carried on by the U.S. Fish Commission Steamer "Albatross" during 1891, Lieut. Commander Z.L. Tanner, U.S.N., Commanding. VII. The Orthoptera of the Galápagos Islands. Bull. Mus. Comp. Zool. Harvard **25**: 1–25, pls. I–III.

Chapter 13
Order Isoptera

The Termites

2 families, 3 genera, 4 species (1 endemic, 3 indigenous)
Minimally 4 natural colonizations

Galápagos termites make hidden nests in the soil or in dead fallen wood or standing trees. They do not make the conspicuous carton or mud nests seen so frequently on trees or fence posts on the Neotropical mainland. They probably colonized by rafting in floating driftwood. All Galápagos species seem to be salt tolerant and three species have been found in moist wood at or near the sea coast. Galápagos *Incisitermes* are most abundant in dead mangrove and *Bersera* trees in the littoral and arid zones, and may also infest buildings. Additional species may have been brought by human commerce, but are not yet known. Winged adults are most frequently collected at night at lights after a heavy rain at the start of the rainy season. Dry-wood termites are collected by chopping and sawing infested wood into short lengths and splitting it carefully to remove winged adults, workers and soldiers. The easiest identifications are made with soldiers, which should be collected into alcohol (keeping each colony collection separate). Constantino (1998) is a catalogue of the 86 genera and 543 species of New World termites.

Key to Galápagos termites

1a Head of soldiers longer than broad; mandibles of soldiers long and thin, without prominent marginal teeth; tibial spurs 3:2:2; soldiers with clypeal elongation over base of mandibles; winged adults with fontanelle (small pale spot or pit in middle of head between eyes); wings at midpoint with only two heavy veins on anterior margin; Rhinotermitidae
. . . . *Heterotermes convexinotatus*

1b Head of soldiers long, or short and anteriorly hollowed out; mandibles of soldiers with prominent marginal teeth; tibial spurs 3:3:3; winged adults without fontanelle; wings at midpoint with three or more heavy veins in anterior margin; Kalotermididae. 2

2a Soldiers with third antennal segment longer than adjoining segments; head not with projecting ridge (brow) forming cavity above mouthparts; front wing with median vein running about midway between radial sector and cubitus (see figure in Isoptera chapter in Borror et al. (1989)) . . . . *Incisitermes pacificus*

2b Soldiers with third antennal segment small; head with projecting ridge (brow) forming cavity above mouth parts; front wing with median vein running much closer to radial sector or uniting with it; *Cryptotermes* . 3
3a Soldiers with head dorsum wrinkled . . . . *C. darwini*
3b Soldiers with head dorsum smooth. . . . . *C. fatulus*

Family Kalotermitidae
The Drywood Termites

This family is probably pre-adapted for over-seas dispersal. The species live in coastal sites, they are relatively tolerant of sea water, and they nest in dead trees which are ready-formed dispersal rafts which are partially resistant to the entry of seawater. Flight dispersal to the islands is unlikely because the maximum flight range of termites is only a few kimometers (Abe 1987).

Genus *Cryptotermes* Banks 1906

C. darwini (Light) 1935: 242 (*Kalotermes* (*Cryptotermes*)); Bacchus 1987: 47
    Distribution. Endemic; Floreana, Gardner, Isabela (n. coast, Alcedo, Tagus Cove), Santiago, Santa Cruz, Wolf.
    Bionomics. Arid and transition zones; February, May–June. In standing dead trees or in fallen branches. The soldiers have dark, wrinkled frontal cavities in their heads to hold a yellow liquid, a defense against ants. Hickin (1979: 152) indicates this species may be a synonym of *C. brevis* which causes much damage to wooden buildings in Florida, the West Indies, South America, Africa, and Hong Kong.

C. fatulus (Light) 1935: 245 (*Kalotermes* (*Cryptotermes*)), replacement name; Bacchus 1987: 55
    Kalotermes occidentalis Light 1933: 103; Socorro Island (Tres Marias Islands, Mexico)
    Distribution. Indigenous?; Isabela, Santiago (nw slope at 600 m).
    Bionomics. Humid forest, in dead bush, and dead branch of live tree; May. Only collected two times.

Genus *Incisitermes* Krishna 1961

I. pacificus (Banks) 1901: 545 (*Calotermes, Kalotermes*); Light 1935: 237; Krishna 1961: 356.
    I. galapagoensis (Banks) 1901: 544 (*Calotermes, Kalotermes*)
    I. immigrans Snyder 1922: 2–4 (*Kalotermes* s. str.); Light 1935: 236
    Distribution. Indigenous?; El Salvador, Panama, Peru (?; see below); Baltra, Bartolomé, Fernandina, Floreana, Genovesa, Isabela, Marchena, Pinta, Pinzón, Rábida, San Cristóbal, Santa Cruz, Santa Fé, Santiago, Seymour, Wolf.

Bionomics. Littoral, arid, transition, and humid forest, pampa, urban and agriculture zones; dwelling in dead wood; January–June, and December; winged adults at lights; the most frequently collected species.

Notes: Light (1935) states that this species should be considered as a senior synonym of *Incisitermes tabogae* Snyder 1924, but Nickle and Collins (1992) use *I. tabogae* as a valid species. *I. galapagoensis* is a supposed endemic species from Islas Wolf and Genovesa. This species cannot be recognized by the descriptions. Cotypes are in the USNM, but Constantino (1998) states that the holotype adult was in CASC, and was destroyed. The distribution of *I. immigrans* is supposed to be Central Pacific Islands, (Hawaii, Marquesas, Fanning, Jarvis), El Salvador, Panama, and Peru; and widespread in the Galápagos. Light (1935) gives characters of soldiers to separate *I. immigrans* from *I. tabogae* (= *I. pacificus*), but I find both "species" of soldiers in the same colony. I have listed above all island records of *Incisitermes* as *I. pacificus*, since it is the oldest available name for the Galápagos. My specimens are deposited in the USNM, and these may help solve the questions about the correct application of these names.

Family Rhinotermitidae
Genus *Heterotermes* Froggatt 1896

H. convexinotatus (Snyder) 1924: 15
   H. orthognathus Light 1933: 113; 1935: 248
   Distribution. Indigenous?; Mexico, West Indies, Central America, to Venezuela: Eden, Santa Cruz, South Plazas.
   Bionomics. Littoral zone under litter of *Cryptocarpus*, manchineel, and *Sesuvium*; and arid zone near coast in rotted *Opuntia*; January, April–May. These are subterranean termites, with colonies where access to moisture is available. They can be serious pests of wood and structures in contact with soil.
   Notes. Nickle and Collins (1992) list this as *H. aureus convexinotatus* for Panama.

References

Abe, T. 1987. Evolution of life-types in termites. *In* Evolution and coadaptation in biotic communities. *Edited by* S. Kawano, J.H. Connell, and T. Hidaka. University of Tokyo Press, Tokyo. pp. 125–148.

Bacchus, S. 1987. A taxonomic and biometric study of the genus *Cryptotermes* (Isoptera: Kalotermitidae). Trop. Pest. Bull. **7**: 1–91.

Banks, N. 1901. Papers from the Hopkins Stanford Galápagos Expedition, 1898–1899. V. Entomological results (5): Thysanura and Termitidae. Proc. Wash. Acad. Sci. **3**: 541–546.

Borror, D.J., Triplehorn, C.A., and Johnson, N.F. 1989. An introduction to the study of insects. Holt, Reinhart and Winston. 6th ed.

Constantino, R. 1998. Catalog of the living termites of the New World (Insecta: Isoptera). Arq. Zool. (São Paulo) **35**: 135–230.

Hickin, N. 1979. Animal life of the Galápagos. Ferendune Books, Oxon, U.K. 236 pp.

Kirby, H., Jr. 1939. The Templeton Crocker Expedition of the California Academy of Sciences, 1932. no. 39. Two new flagellates from termites in the genera *Coronympha* Kirby, and *Metacoronympha* Kirby, new genus. Proc. Calif. Acad. Sci. ser. 4, **22**: 207–220, 4 pls.

Krishna, K. 1961. A generic revision and phylogenetic study of the family Kalotermitidae (Isoptera). Bull. Am. Mus. Nat. Hist. **122** (4): 303–408.

Light, S.F. 1933. Termites of western Mexico. Univ. Calif. Publ. Entomol. **6**: 79–164, pls. 7–11.

Light, S.F. 1935. The Templeton Crocker Expedition of the California Academy of Sciences, 1932. no. 20. The termites. Proc. Calif. Acad. Sci. ser. 4, **21**: 233–258, pls. 9–10, 10 t. Figures

McNeill, J. 1901. Papers from the Hopkins Stanford Galápagos Expedition, 1898–1899. Entomological Results (4): Orthoptera. Proc. Wash. Acad. Sci. **3**: 487–506.

Nickle, D.A., and Collins, M.S. 1992. The termites of Panama. *In* Insects of Panama and Mesoamerica: Selected Studies. *Edited by* D. Quintero and A. Aiello. Oxford Univ. Press, New York. pp. 208–241.

Snyder, T.E. 1922. New termites from Hawaii, Central and South America, and the Antilles. Proc. U.S. Natl. Mus. **61** (art. 20): 1–32.

Snyder, T.E. 1924. Descriptions of new species and hitherto unknown castes of termites from America and Hawaii. Proc. U.S. Natl. Mus. **64** (art. 6): 1–40.

Snyder, T.E. 1949. Catalog of the termites (Isoptera) of the world. Smithson. Misc. Collect. **112**: 1–490.

Snyder, T.E. 1954. Order Isoptera. The termites of the United States and Canada. Tech. Bull. Rational Pest Control Assoc. New York. 64 pp.

Chapter 14
Order Dermaptera

The Earwigs

2 families, 5 genera, 7 species (2 endemic, 1 indigenous, 4 adventive)
Minimally 2 natural colonizations

Earwigs are predominantly a tropical group, with about 1800 species worldwide. Over 300 species occur in the Neotropical region, mostly in the family Labiidae (Brindle 1971a; Reichardt 1968–1971; Steinmann 1973, 1975). They are elongate, somewhat flattened insects, with characteristic forceps at the end of the abdomen. Most species are nocturnal, and feed mainly as predators on insects or as scavengers on dead plant materials, but some may feed on live plant tissue. Dermaptera are usually poorly represented in island faunas (Brindle 1972). Three species in the Galápagos are indigenous to Europe and have probably been transported by humans. Only these three adventive Galápagos species have functional wings as adults. Other species may be found in the future to be adventive from tropical America (see key in Hoffman 1987). *Anisolabis maritima* occurs in the littoral zone and may have colonized by rafting. The ancestor of two species of small, eyeless, soil-dwelling *Anopthalmolabis* probably naturally colonized the Galápagos in an eyeless condition by rafting in soil from the South American continent.

The abdomens of males and females of each species are sexually dimorphic in that they have forceps (modified cerci) of different shapes. Adult males usually have distinctly curved forceps and females have more or less straight forceps. Immatures have fewer antennal segments than adults, and are distinguished from adults by the combination of a male-like 10-segmented abdomen (adult females have only 8 apparent segments) with female-like straight forceps.

Key to Galápagos Dermaptera

1a Second tarsal segment typically cylindrical, not projecting beneath third; third segment (in lateral view) arising more or less terminally from second toward ventral part; Superfamily Labioidea 2
1b Second tarsal segment typically flattened, projecting distinctly beneath third (or if not, then third (in lateral view) arising subterminally toward dorsal part of second segment) (likely to be adventive; not yet known from Galápagos) . . . . Superfamily Forficuloidea
2a Tegmina (meso-thoracic wings) absent, or present only as rounded flaps that do not meet at inner basal margins, hind wings absent; body

large or small; male forceps asymmetrical, right curved inward more
than left; Carcinophoridae. 3

2b Tegmina present, hind wings present; body smaller; male forceps symmetrical; Labiidae 6

3a Eyeless; body smaller (adults under 6 mm long); *Anophthalmolabis* . . 4

3b Fully eyed; body larger (adults over 10 mm long) 5

4a On Isla Santa Cruz, arid zone, in litter or in deep lava crevices
. . . . *A. leleupi*

4b On Isla Isabela, paramo zone, in litter or soil . . . . *A. n. sp.*

5a Antennae with 20–24 segments, with basal segment longer than combined segments 4–6; legs uniformly pale yellow; body length with forceps 20–25 mm; forceps as in Figs. 14.1, 14.2 . . . *Anisolabis maritima*

5b Antennae with 14–18 segments, with one or more distal segments
white, with basal segment about equal to combined segments 4–6; legs
pale, but with femora and tibiae usually distinctly (but sometimes
faintly) banded with black; body length with forceps 12–20 mm; forceps as in Figs. 14.3, 14.4 . . . . *Euborellia annulipes*

6a Body covered with brownish or yellowish pubescence; hind wings visible when folded, twice as long as tegmina; antenna filiform, with
fourth segment cylindrical, as long as third segment 7

6b Body not pubescent; tegmina short, with hind wings absent or concealed; antennae clavate, with fourth segment conical, shorter than
third segment (likely to be adventive; not yet known from Galápagos)
. . . . *Marava*

7a Male forceps with either broad or narrow tooth on inner surface near
base; female pygidium not as below 8

7b Male forceps without tooth on inner surface (Fig. 14.9); female pygidium
about as long as broad, postero-lateral projections slightly divergent,
occupying almost entire distal margin. . . . . *Labia annulata*

8a Forceps of male curved, with subbasal notch and narrow tooth, apical
inner margin smooth; forceps of female slender and gently curved; forceps as in Figs. 14.5, 14.6 . . . . *Circolabia arcuata*

8b Forceps of both sexes very broad at base; forceps of male strongly
curved, inner margin serrate; tooth broad; forceps of female stout; forceps as in Figs. 14.7, 14.8 . . . . *Labia curvicauda*

160

Figs. 14.1–14.9. Forceps of Galápagos earwigs. Fig. 14.1. *Anisolabis maritima*, male. Fig. 14.2. *Anisolabis maritima*, female. Fig. 14.3. *Euborellia annulipes*, male. Fig. 14.4. *Euborellia annulipes*, female. Fig. 14.5. *Circolabia arcuata*, male. Fig. 14.6. *Circolabia arcuata*, female. Fig. 14.7. *Labia curvicauda*, male. Fig. 14.8. *Labia curvicauda*, female. Fig. 1.4.9. *Labia annulata*, male.

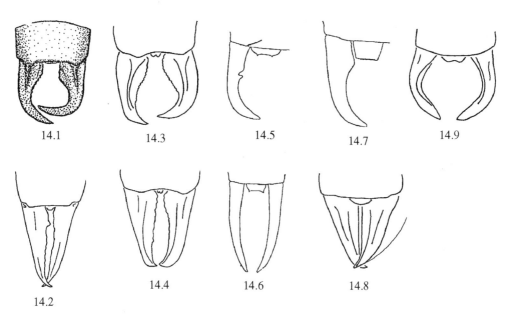

14.1   14.3   14.5   14.7   14.9

14.4   14.6   14.8

14.2

Family Carcinophoridae
Genus *Anisolabis* Fieber 1853

A. maritima (Bonelli) 1832: 221; Brindle 1969; 333. The seaside earwig.
   Distribution. Possibly indigenous; cosmopolitan along tropical and temperate sea coasts; Floreana, San Cristóbal, Santiago
   Bionomics. Littoral zone, living beneath debris on beaches, along seashores and in mangroves; sometimes in buildings; predatory on smaller insects; March; not commonly found – perhaps not established; without functional wings.

Genus *Euborellia* Burr 1909

E. annulipes (Lucas) 1847: 84 (*Forficesila*); McNeill 1901: 492; Brindle 1969: 333; Parkin et. al. 1972: 101. The ring-legged earwig.
   Anisolabis bormansi Scudder 1893: 5.
   Distribution. Probably adventive; cosmopolitan; Floreana, Isabela, San Cristóbal, Santa Cruz, Santiago.

Bionomics. Littoral to pampa and agricultural zones; January–December; scavenger in litter and debris; often taken in pit traps; without functional wings. This is the most common Galápagos species, especially in dark and humid habitats. It can cause minor damage to plants and stored foods, and will attack other insects. I judge it to be adventive because it is widespread on only the five islands most influenced by humans.

Genus *Anopthalmolabis* Brindle 1968

A. leleupi Brindle 1968: 172; 1969: 333.
  Distribution. Endemic; Santa Cruz.
  Bionomics. Arid zone; February, May, December. A small (5 mm long body) eyeless, wingless species from litter in deep crevices (Grieta Iguana, CDRS). Other than the two Galápagos species, the genus has a larger eyeless species, *A. caeca* Borelli (9 mm long), in Santa Fé Prov., Argentina (Brindle 1980), which was formerly placed in the genus *Anisolabis*. The only other eyeless earwigs in the world are in Hawaii, Reunion Island (Brindle 1980), and the Canary islands, all of which are young volcanic oceanic islands.

A. new species
  Distribution. Endemic; Isabela.
  Bionomics. Pampa zone; July. An eyeless, wingless species known from one specimen from under moss mats and rocks on Sierra Negra, about 640 m (Campamento de los Pumas). This population is thought to be a separate species from that on Santa Cruz because it seems most likely that these two eyeless island populations are reproductively isolated from each other by the sea.

Family Labiidae
The Little Earwigs
Genus *Circolabia* Steinmann 1987

C. arcuata (Scudder) 1876: 257 (*Labia*); Brindle 1969: 333 (*Labia*).
  Distribution. Adventive?; South America, West Indies, Central America, Mexico (Sakai 1970); Floreana, San Cristóbal, Santa Cruz.
  Bionomics. Scavenger and maybe a predator; in litter and debris; arid to pampa and agricultural zones; February–June, December; with functional wings. Brindle (1969: 334) notes that he thought that this species (as *Labia arcuata*) may not be a significant record for the Galápagos, and that it is a "casual." However, it is now widespread on three islands, and occurs in native habitats.

Labia Leach 1815

L. annulata (Fabricius) 1793.

Distribution. Adventive; West Indies and Panama (Brindle 1971*b*); *San Cristóbal* (El Junco). This is a new record for the Galápagos Archipelago.

Bionomics. Scavenger or predator; pampa zone; in leaf litter; February; with fully developed wings.

L. curvicauda (Motschulsky) 1863: 2 (*Forficesila*).

Distribution. Adventive; cosmopolitan (Brindle 1971*b*, 1972); Santa Cruz. This is a new record for the Galápagos Archipelago.

Bionomics. Scavenger or predator; arid zone; collected in rotted *Opuntia* trunk at CDRS in May; with fully developed wings.

See Brindle (1971*a*) for key to Neotropical species of *Labia*. Note that Helfer (1963: 17) notes *Labia minor* (Linnaeus), the cosmopolitan least earwig, from the Galápagos. There are no records to verify this. I think the record may represent *C. arcuata*.

Acknowledgements

Fabian Haas (Universität Ulm, Ulm, Germany) helped with identifications and literature.

References

Brindle, A. 1968. A new genus and species of blind Dermaptera from the Galápagos Islands. Mission zoologique belge aux iles Galápagos et en Ecuador (N. et J. Leleup, 1964–1965). Resultats scientifiques **1**: 171–176, 2 Figures.

Brindle, A. 1969. The earwigs (Dermaptera) of the Galápagos Islands. Entomol. Rec. **81**: 331–334, 1 map.

Brindle, A. 1971*a*. A revision of the Labiidae (Dermaptera) of the Neotropical and Nearctic Regions. J. Nat. Hist. **5**: 155–182.

Brindle, A. 1971*b*. Bredin-Archbold-Smithsonian Biological Survey of Dominica: the Dermaptera (earwigs) of Dominica. Smithson. Contrib. Zool. **63**. 24 pp.

Brindle, A. 1972. Insects of Micronesia: Dermaptera. Insects of Micronesia **5**: 91–171.

Brindle, A. 1980. The cavernicolous fauna of Hawaiian lava tubes. 12. A new species of blind troglobitic earwig (Dermaptera, Carcinophoridae), with a revision of the related surface-living earwings of the Hawaiian Islands. Pac. Insects **21**: 261–274.

Hebard, M. 1920. Expedition of the California Academy of Sciences to the Galápagos Islands, 1905–1906. XVII. Dermaptera and Orthoptera. Proc. California Acad. Sci. ser. 4, **2**: 311–346, pl. 18, t. Figs. 1–11.

Helfer, J.R. 1963. How to know the grasshoppers, cockroaches, and their allies. Wm. C. Brown, Dubuque, IA. 359 pp.

Hoffman, K.M. 1987. Earwigs (Dermaptera) of South Carolina, with a key to the eastern North American species and a checklist of the North American fauna. Proc. Entomol. Soc. Wash. **89**: 1–14.

McNeill, J. 1901. Papers from the Hopkins Stanford Galápagos Expedition, 1898–1899. Entomological Results (4): Orthoptera. Proc. Wash. Acad. Sci. **3**: 487–506.

Parkin, P., Parkin D.T., Ewing, A.W. and Ford, H.A. 1972. A Report on the Arthropods Collected by the Edinburgh University Galápagos Islands Expedition, 1968. The Pan-Pacific Entomologist **48**: 100-107.

Reichardt, H. 1968–1971. Catalogue of New World Dermaptera (Insecta). Dept. Zool. Sec. Agric. (Sao Paulo), Pap. Avul. Zool. **21**: 183–193 (Pt. I. Introduction and Pygidicranoidea); **22**: 35–46 (Pt. II. Labioidea, Carcinophoridae); **23**: 83–109 (Pt. III. Labioidea, labidae); **24**: 161–184 (Pt. IV. Forficuloidei); **24**: 221–257 (Pt. V. Additions, corrections, bibliography and index).

Sakai, H. 1970. Dermapterorum Catalogus Praeliminaris. II. A basic survey for numerical taxonomy of the Labiidae, the Dermaptera of the world. Daito Bunka University, Tokyo, Japan.

Scudder, S.H. 1893. Reports on the dredging operations off the west coast of Central America to the Galápagos, to the west coast of Mexico, and in the Gulf of California, in charge of Alexander Agassiz, carried on by the U.S. Fish Commission Steamer "Albatross" during 1891, Lieut. Commander Z.L. Tanner, U.S.N., Commanding. VII. The Orthoptera of the Galápagos Islands. Bull. Mus. Comp. Zool. Harvard **25**: 1–25, pls. I–III.

Steinmann, H. 1973. A zoogeographical checklist of world Dermaptera. Fol. Entomol. Hungarica **26**: 145–154.

Steinmann, H. 1975. Suprageneric classification of Dermaptera. Acad. Sci. Hungaricae Acta Zool. **21**: 195–220.

Steinmann, H. 1987. Two new genera and species for the subfamily Labiinae (Dermaptera: Labiidae). Acta Zool. Hungarica **33**: 177–186.

Chapter 15
Order Embioptera

The Webspinners

2 families, 2 genera, 2 species (1 endemic, 1 adventive),
Minimally 1 natural colonization

Embiids are strange elongate insects that live gregariously in silken tubes or galleries on or under rocks, on tree trunks, and in vegetation. The silk is spun from glands on the males' front tarsi. The males have wings and can fly (Ross 1970, 1984) but mature females are always wingless so dispersal of every species is limited. The endemic Galápagos species may have colonized by transport in the feathers of birds, and the second species was introduced by commerce. Rafting transport of colonies is possible, but seems less likely to account for the Galápagos species.

Key to Galápagos Embiidina

1a Adult males (with wings) with segments of left cercus fused together; Anisembiidae . . . . *Chelicerca*
1b Adult males with segments of left cercus clearly two segmented; Oligotomidae . . . . *Oligotoma*

Family Anisembiidae
Genus *Chelicerca* Ross 1940

C. galapagensis Ross 1966: 501.
   Distribution. Endemic; Pinzón, San Cristóbal, Santa Cruz, Wolf.
   Bionomics. Arid to humid forest zones; February–July. Feeding on lichens, or detritus. The silk galleries may occur on bushes and tree trunks, on lichen covered rocks, and in sea bird rookeries. Ross (1966) suggests that the ancestors dispersed to the islands in the plumage of birds. It is closely related to species of coastal arid Ecuador and Peru.

Family Oligotomidae
Genus *Oligotoma* Westwood 1836

O. saundersii (Westwood) 1837: 373.
   Distribution. Adventive; tropicopolitan; San Cristóbal, Santa Cruz.
   Bionomics. Males were first found at lights at night in Puerto Ayora in 1991; colonies were then found in silken webs in litter of forests on farms at

Bellavista, and large webs with pronounced silken tunnels were on tree trunks and branches in arid zone vegetation along paths at the Darwin Station in 1999; March–June. This species originated in India and has been spread widely through the tropics and subtropics by commerce.

References

Ross, E.S. 1966. A new species of Embioptera from the Galápagos Islands. Proc. Calif. Acad. Sci., ser. 4, **34**: 499–501, 1 Figure.

Ross, E.S. 1970. Biosystematics of the Embioptera. Ann. Rev. Entomol. **15**: 157–172.

Ross, E.S. 1984. A synopsis of the Embiidina of the United States. Proc. Entomol. Soc. Wash. **86**: 82–93.

Chapter 16
Order Zoraptera

The Zorapterans

1 family, 1 genus, 1 species (indigenous?)
Minimally 1 natural colonization

Zorapterans are small, pale, gregarious insects living in rotting logs. The adult males and females have a short flight before sheding their wings and establishing a colony. Dispersal to the Galápagos was probably by rafting, because flight across the oceanic barrier is unlikely.

Family Zorotypidae

Genus *Meridozoros* Kukalova-Peck and Peck 1993

M. leleupi Weidner 1976: 162 (*Zorotypus*); Kukalova-Peck and Peck 1993: 341 (description of wings).

    Distribution. Indigenous?, Venezuela?; Isabela, Santa Cruz.

    Bionomics. Scavenger on fungus spores or small dead arthropods?; humid forest; February–May. Under tree bark; in wet forest litter, also taken in flight intercept traps in humid forest (winged males and females, Kukalova-Peck and Peck 1993). A specimen from Venezuela seems to be this species, but the identification needs to be confirmed. Another species, *Latinozoros barberi* (Choe), occurs on Cocos Island, and in Panama and the Dominican Republic (Kukalova-Peck and Peck 1993).

References

Kukalova-Peck, J., and Peck, S.B. 1993. Zoraptera wing structures: evidence for new genera and relationship with the blattoid orders (Insects: Blattoneoptera). Syst. Entomol. **18**: 333–350.

Weidner, M. 1976. XXIII. Eine neue *Zorotypus*-Art von den Galápagosinseln, *Zorotypus leleupi* species n. (Zoraptera). Mission zoologique belge aux iles Galápagos en Ecuador (N. et J. Leleup; 1964–1965) **3**: 161–176.

Chapter 17
Order Psocoptera

The Bark Lice

15 families, 22 genera, 40 species
(20 endemic, 6 indigenous, 14 introductions)
Minimally 20 natural colonizations

Galápagos Psocopterans are unusually diverse, abundant, and dominant in soils and litter as well as on vegetation in all life zones, including the arid zone. Generally, they feed as scavengers, and in the Galápagos they may help fill the litter decomposer role usually occupied elsewhere by collembola, because many may better tolerate the arid conditions of the lowlands of the islands. Alternatively, especially in the humid zones, they may be specialist consumers on lichens, molds, or microepiphytic (protococcal) algae and microepiphytic fungi on leaves, bark and branches of shrubs and trees (Thornton 1984). They are inconspicuous insects except for the large sheet webs of *Archipsocus* which may cover large areas of tree trunks or branches in the arid zone. Mockford (1993) gives additional biological information on Psocoptera. Almost all Galápagos collections were made by inspecting plant surfaces, or beating arboreal vegetation. Dispersal was probably mostly by winged adults in aerial plankton, and by some rafting of adults and nymphs. Many psocopterans seem to be parthenogenetic, but information is not available on the Galápagos species. The only comprehensive reference is Thornton and Woo (1973), resulting from 6 months of field study in the islands by Thornton in 1967. The affinities of the fauna are with: 1) South America, 2) Central America, 3) the Pacific region, and 4) North America. Comparisons of the Hawaiian and Galápagos faunas are in Thornton (1984), and of other Pacific islands in Thornton (1989). My extensive collections are with E. Mockford, but have not yet been studied.

Suborder Trogiomorpha
Family Lepidopsocidae
Genus *Echmepteryx* Aaron 1886

E. (Loxopholia) aperta Thornton and Woo 1973: 7.
  Distribution. Endemic; Floreana, Isabela, Santa Cruz, San Cristóbal
  Bionomics. Scavenger; transition to pampa and agriculture zones; on many types of plants; March–May.
  Relationships. Similar to *E. lealea* found in bird nests in Brazil.

E. (Thylacopsis) lunulata Thornton, Lee and Chui 1972: 64; Thornton and Woo 1973: 6.
  Distribution. Indigenous; Micronesia and Hawaii; Floreana, Isabela, Pinta, Santiago, Santa Cruz, San Cristóbal.
  Bionomics. Scavenger; arid to pampa and agriculture zones; occurs on many plant species; January–June.

E. (Thyacopsis) madagascariensis (Kolbe) 1885: 184 (*Thylax*); Thornton and Woo 1973: 7.
  Distribution. Adventive; circum-Pacific and trans-Pacific; Africa; Floreana, Isabela, Santa Cruz.
  Bionomics. Scavenger; found on many plant species in the agricultural zone; usually an inhabitant of dead persistent leaves; March–May.

Genus *Lepidopsocus* Enderlein 1963

L. maculatus Thornton, Lee and Chui 1972: 68; Thornton and Woo 1973: 8.
  Distribution. Indigenous?; Micronesia and Hawaii; Fernandina, Floreana, Genovesa, Isabela, Pinzón, San Cristóbal, Santa Cruz, Santiago.
  Bionomics. Scavenger; littoral to humid forest and agriculture zones; on many types of plants; February–May.

Genus *Nepticulomima* Enderlein 1906

N. cavagnaroi Thornton and Woo 1973: 9.
  Distribution. Endemic; Santa Cruz.
  Bionomics. Scavenger; humid forest zone (Horneman Farm); May.

Genus *Soa* Enderlein 1904

S. reticulata Thornton and Woo 1973: 11.
  Distribution. Endemic; Santa Cruz.
  Bionomics. Scavenger; humid forest zone (Horneman Farm); May.
  Notes. Similar to *S. flaviterminata* of Africa, Asia, and South America.

Family Trogiidae
Genus *Cerobasis* Kolbe 1882

C. lambda Thornton and Woo 1973: 16.
  Distribution. Endemic; Pinzón, San Cristóbal, Santiago.
  Bionomics. Scavenger; littoral to pampa zone; on many kinds of plants; February, April.
  Notes. Most similar to *C. guestfalica* of Argentina.

C. recta Thornton and Woo 1973: 17.
  Distribution. Endemic; Santiago.

Bionomics. Scavenger; pampa zone; known only from dead ferns hanging over water at 490 m; April.

C. treptica Thornton and Woo 1973: 12.
  Distribution. Endemic; widespread: Darwin, Española, Fernandina, Floreana, Genovesa, Isabela, Marchena, Pinta, Pinzón, Rábida, Seymour, San Cristóbal, Santa Cruz, Santa Fé, Santiago, Tortuga, Wolf.
  Bionomics. Scavenger; littoral to pampa zone, most abundant in lowlands, on many types of vegetation; three forms are known with differences in pigmentation; March–June.

Family Psoquillidae
Genus *Psoquilla* Hagen 1865

P. marginepunctata Hagen 1865: 123; Thornton and Woo 1873: 19.
  Distribution. Adventive; tropicopolitan; Santa Cruz.
  Bionomics. Scavenger; agriculture zone, on Horneman Farm; associated with stored products and on and under tree bark; May.

Genus *Rhyopsocus* Hagen 1876

R. orthatus Thornton and Woo 1973: 19.
  Distribution. Endemic; Santa Cruz.
  Bionomics. Scavenger; agricultural zone; May.

Suborder Troctomorpha
Family Liposcelidae
Genus *Embidopsocus* Hagen 1866

E. pauliani Badonnel 1955: 78; Thornton and Woo 1973: 21.
  Distribution. Adventive?; Africa; San Cristóbal (base of Cerro Brujo).
  Bionomics. Scavenger; pampa zone; May.

E. thorntoni Badonnel 1971: 325.
  Distribution. Endemic; Santa Cruz.
  Bionomics. Scavenger; agriculture zone; May.
  Notes. The species shows "some slight similarity" with *E. flexuosus* from Brazil and Argentina.

Genus *Liposcelis* Motschulsky 1852

L. entomophila (Enderlein) 1907: 34 (*Troctes*); Thornton and Woo 1973: 21.
  Distribution. Adventive; cosmopolitan; Pinzón (caldera area); February.
  Bionomics. Scavenger; humid zone; the species is widely distributed by commerce, and occurs in stored grain. It should be found around buildings on the inhabited islands.

Family Pachytroctidae
Genus *Tapinella* Enderlein 1908

T. francesca Thornton and Woo 1973: 22.
   Distribution. Endemic; Pinta, San Cristóbal (Cerro Brujo).
   Bionomics. Scavenger; in humid forest and *Zanthoxylum* zones; May and June.

T. stenomedia Thornton and Woo 1973: 23.
   Distribution. Endemic; Santa Cruz.
   Bionomics. Scavenger; agricultural zone; May.

Genus *Pachytroctes* Enderlein 1905

P. (*Neotroctes*) *achrosta* Thornton and Woo 1973: 24.
   Distribution. Endemic; Santa Cruz.
   Bionomics. Scavenger; arid and agricultural zones; in humus; apterous;
      January, May.

Suborder Psocomorpha
Family Epipsocidae
Genus *Epipsocus* Hagen 1866

E. campanulatus Thornton and Woo 1973: 25.
   Distribution. Endemic; Santa Cruz.
   Bionomics. Scavenger; agricultural zone; May.

Family Caeciliidae
Genus *Caecilius* Curtis 1837

C. antillanus Banks 1938: 288; Thornton and Woo 1973: 27.
   Distribution. Indigenous?; Florida, Mexico, and West Indies to NE South
      America; Floreana, San Cristóbal.
   Bionomics. Scavenger; from arid to agriculture zone; no males were found,
      perhaps parthenogenetic; March–May.

C. distinctus Mockford 1966: 137; Thornton and Woo 1973: 27.
   Distribution. Indigenous?; Mexico, Surinam, and Peru; San Cristóbal.
   Bionomics. Scavenger; at 360 m on *Psidium guajava*; February.

C. insularum Mockford 1966: 157; Thornton and Woo 1973: 27.
   Distribution. Indigenous?; Florida, Mexico, and and West Indies to north
      coast of South America; Isabela, Santa Cruz, Santiago.
   Bionomics. Scavenger; agricultural zone; on fruit trees; March–May.

Family Lachesillidae
Genus *Lachesilla* Westwood 1840

L. aethiopica Enderlein 1902: 11; Thornton and Woo 1973: 28.
  Distribution. Adventive?; Africa, West Indies, Central America, South America; Floreana, Santa Cruz, Santiago.
  Bionomics. Scavenger; on adventive plants in agricultural zone, and at 490 m in ferns on Santiago; perhaps parthenogenetic; March–May.

L. castroi Thornton and Woo 1973: 28.
  Distribution. Endemic; Fernandina, Floreana, Genovesa, Marchena, Pinta, Pinzón, Rábida, San Cristóbal, Santa Cruz, Santiago, Tortuga, Wolf.
  Bionomics. Scavenger; littoral to pampa zone; on various dry plant materials; March–June.

Family Ectopsocidae
Genus *Ectopsocus* McLachlan 1899

E. maindroni Badonnel 1935: 81; Thornton and Woo 1973: 31.
  Distribution. Adventive; cosmopolitan; Genovesa, Isabela, San Cristóbal, Santa Cruz.
  Bionomics. Scavenger; littoral and arid zone, and an associate with man's dwellings and produce; all Galápagos collections are from near inhabited areas except for on Genovesa; February–May.

E. meridionalis Ribaga 1904: 294; Thronton and Woo 1973: 31.
  Distribution. Adventive; cosmopolitan; Isabela, Pinta, Santa Cruz, Santiago.
  Bionomics. Scavenger; from transition to pampa zone (1570 m on Volcán Wolf), especially on plants in agriculture zone; probably parthenogenetic; February–June.

E. richardsi (Pearman) 1929: 105; Thornton and Woo 1973: 32.
  Distribution. Adventive; cosmopolitan; Santa Cruz.
  Bionomics. Scavenger; a stored products associate; the single Galápagos record was from an upland farm; no month data.

E. species, Thornton and Woo 1973: 32; to be described by S.K. Wong.
  Distribution. Endemic?; Fernandina, Floreana, Gardner near Hood, Genovesa, Isabela (Sierra Negra, V. Darwin), Seymour Norte, Santiago, Pinta, Pinzón, Plaza Sur, San Cristóbal, Santa Cruz, Santiago, Wolf.
  Bionomics. Scavenger. No other data.

Family Peripsocidae
Genus *Peripsocus* Hagen 1866

P. pauliani Badonnel 1949: 42; Thornton and Woo 1973: 32.
> Distribution. Adventive; cosmopolitan; Española, Fernandina, Floreana, Genovesa, Isabela, Marchena, Pinta, Pinzón, Rábida, San Cristóbal, Santa Cruz, Santiago, Tortuga, Wolf.
> Bionomics. Scavenger; arid to pampa zone; found on many species of plants; probably parthenogenetic; March–June.

P. potosi Mockford 1971: 110; Thornton and Woo 1973: 33.
> Distribution. Indigenous?; Mexico, West Indies, and Central America; Floreana, Isabela, Pinta, San Cristóbal, Santa Cruz, Santiago.
> Bionomics. Scavenger; found from transition to pampa zone on many plant species; probably parthenogenetic; March–June.

P. species, Thornton and Woo 1973: 33; to be described by S.K. Wong.
> Distribution. Endemic; Isabela (Sierra Negra, Cerro Azul).
> Bionomics. Scavenger; only at 500 m and above, no other data.

P. stagnivagus Chapman 1930: 376; Thornton and Woo 1973: 33.
> Distribution. Adventive?; USA; Fernandina, Floreana, Genovesa, Isabela, Marchena, Pinta, Pinzón, Rábida, San Cristóbal, Santa Cruz, Santa Fé, Santiago, Tortuga.
> Bionomics. Scavenger; arid to pampa zone (1570 m on V. Wolf); on many plant species; February–June.

Family Pseudocaeciliidae
Genus *Pseudocaecilius* Enderlein 1903

P. criniger (Perkins) 1899: 85; Thornton and Woo 1973: 34.
> Distribution. Adventive; cosmopolitan; Española, Floreana, Isabela, Pinta, Santa Cruz, Santiago.
> Bionomics. Scavenger; transition to evergreen shrub zone; March–May.

P. tahitiensis (Karny) 1926: 288; Thornton and Woo 1973: 34.
> Distribution. Adventive?; Micronesia; Tahiti; Isabela, Rábida, San Cristóbal, Santa Cruz, Santiago, Seymour.
> Bionomics. Scavenger; arid to pampa zone; on foliage of many plant species; March–May.
> Notes. Probably a sister species of the above species.

Family Archipsocidae
Genus *Archipsocus* Hagen 1882

A. spinosus Thornton, Lee and Chui 1972: 126 (subgenus *Archipsocus*); Thornton and Woo 1973: 35 (subgenus *Archipsocus*).

Distribution. Adventive?; Micronesia; Isabela, Pinta, San Cristóbal, Santa Cruz, Santa Fé, Tortuga.

Bionomics. Scavenger; arid to humid zones; on many plants; possibly parthenogenetic; March–June. The sheet webs of these insects were common on tree trunks and branches, especially *Parkinsonia* trees, in the arid zone vegetation behind the Darwin Station in 1999. Spines or irregularities hold the webs away from the tree trunk, and the insects can be seen scuttling under the web.

Family Philotarsidae
Genus *Aaroniella* Mockford 1951

A. galapagensis Thornton and Woo 1973: 35.

Distribution. Endemic; Española, Floreana, Isabela, San Cristóbal, Santiago, Santa Cruz.

Bionomics. Scavenger; arid to pampa zone; on many plant species; February–April.

Family Psocidae
Genus *Blaste* Kolbe 1883

B. uncinata Thornton and Woo 1973: 38.

Distribution. Endemic; Floreana, Isabela, Santa Cruz, Santiago, Seymour.

Bionomics. Scavenger; arid to humid zones; on many plant species; March–May.

Genus *Indiopsocus* Mockford 1974

I. acraea (Thornton and Woo) 1973: 41 (*Ptycta*); Thornton 1985: 183.

Distribution. Endemic; Floreana, Isabela, Pinzón, San Cristóbal, Santiago.

Bionomics. Scavenger; in humid and pampa zones; from 500 to 1500 m; March–May.

Notes. This and the next species have differentiated populations which show that they are actively speciating. Related to *I. expansus* of the Columbian Andes at 2750 m (Thornton 1985: 183).

I. dentata (Thornton and Woo) 1973: 44 (*Ptycta*); Thornton 1985: 183.
  Distribution. Endemic; Fernandina, Floreana, Genovesa, Isabela, Marchena, North Seymour, Pinta, Pinzón, Rábida, San Cristóbal, Santa Cruz, Santa Fé, Santiago, Tortuga.
  Bionomics. Scavenger; arid to pampa zones; on many plant species; January–May.
  Notes. Related to species from the Caribbean, the gulf coast of Mexico, and Southern Florida (Thornton 1985: 183).

I. marta (Thornton and Woo) 1973: 47 (*Ptycta*).
  Distribution. Endemic; Floreana, San Cristóbal, Santa Cruz, Wolf.
  Bionomics. Scavenger; arid to humid forest zones; on *Scalesia, Croton, Cordia*, and *Prosopis*; January, May–June.

Family Myopsocidae
Genus *Lichenomima* Enderlein 1910

L. chelata (Thornton and Woo) 1973: 50 (*Myopsocus*); Mockford 1983: 217.
  Distribution. Endemic; Española, Floreana, Isabela, San Cristóbal, Santa Cruz, Santa Fé, Santiago.
  Bionomics. Scavenger; arid to pampa zone (1420 m on V. Wolf); on many plant species; February–May.

Acknowledgements

E.L. Mockford (Illinois State University, Normal, IL) checked this chapter and helped update nomenclature changes.

References

Badonnel, A. 1971. *Embidopsocus thorntoni* (Psocoptera, Liposcelidae) nouvelle espèce de l'archipel des Galápagos. Nouv. Rev. Entomol. **1**: 325–327.

Chiu, V.W.D., and Thornton, I.W.B. 1972. A numerical taxonomic study of the endemic *Ptycta* species of the Hawaiian Islands (Psocoptera: Psocidae). System. Zool. **21**: 7–22.

Mockford, E.L. 1983. Redescription of the type species of *Myopsocus*, *M. undulosus* (Hagen), and resulting nomenclatural changes in genera and species of Myopsocidae (Psocoptera). Psyche J. Entomol. **89**: 211–220.

Mockford, E.L. 1993. North American Psocoptera (Insecta). Flora and Fauna Handbook no.10. Sandhill Crane Press, Gainesville, FL.

Thornton, I.W.B. 1984. Psocoptera of the Hawaiian Islands. Part III. The endemic *Ptycta* complex (Psocidae): systematics, distribution, and evolution. Int. J. Entomol. **26**: 1–128.

Thornton, I.W.B. 1985. The geographical and ecological distribution of arboreal *Psocoptera*. Ann. Rev. Entomol. **30**: 175–196.

Thornton, I.W.B. 1989. Psocoptera (Insecta) of the island of Moorea, French Polynesia, and comparisons with other Pacific Island Faunas. Bull. Mus. Natl. Hist. Nat., Paris, 4 ser; 11, sec. A: 783–828.

Thornton, I.W.B., and Woo, A.K.T. 1973. Psocoptera of the Galápagos Islands. Pac. Insects **15**: 1–58, Figs. 1–96, tables 1–15.

Chapter 18
Order Thysanoptera

The Thrips

4 families, 31 genera, 50 species
(0 endemic?, 45 indigenous?, 8 adventive?),
Minimally 45 natural colonizations

In the only previous work on the thrips of the Galápagos, Bailey (1967) recorded three families, six genera, and eight species. Now, 50 species are known, of which 33 are described and nameable here. The Aeolothripidae is represented by one predaceous species; Merothripidae with one fungivorous species; Thripidae with 18 phytophagous species and 3 predaceous species; and the Phlaeothripidae with seven known phytophagous species and 19 species that feed on fungal hyphae or spores, on break-down products associated with fungal infestations, or prey on other insects.

The Galápagos fauna can be compared to that of Bermuda, another isolated island group located approximately 1050 km east of Cape Hatteras, North Carolina. The faunas of both are similar in having (1) a greater proportion of predatory species than in a continental fauna, (2) many genera represented by one species each, and (3) most or all species are adventive (Nakahara and Hilburn 1989: 251). However, the Galápagos fauna consists of a greater percentage of genera with multiple species, and has about 50% phytophagous species in contrast to 70% from Bermuda.

The namable species are mainly adventive from or indigenous to Central and South America and the Caribbean area. Several are from the Nearctic Region or are widespread in the tropics. Most of the unnamed species probably migrated from Central or South America where the thrips fauna is very poorly known. The natural modes of colonization are by air currents or rafting. Thrips have been recorded from both the air and sea surface between islands (Peck 1994a, 1994b). In the rather short time span that humans have colonized the islands, their activities have undoubtedly contributed to introductions of exotic species of thrips on rooted plants or on cuttings. However, several widespread pest species dispersed mainly by humans, such as *Thrips tabaci* Lindeman, were not collected. Although more than 100 endemic species of *Frankliniella* are known from the New World (Nakahara 1997), only four species have been collected in the Galápagos thus far.

This study collected a predominance of macropterae (fully winged forms), even in phlaeothripid taxa that normally consist mostly of brachypterae (short-winged forms) or apterae (wingless forms). The abundance of macropterae is apparently due to the method of collecting. Most thrips were collected with malaise traps, FIT (flight intercept traps), and pan traps which select for flying insects. But the abundance of macropterae may be real. Mound and O'Neill (1974: 497) found, in island populations of *Merothrips floridensis* Morgan, a predominance of macropterae rather than apterae. The opposite condition, a prevalence of apterae, is the normal situation for specimens from the continental United States. They also noted this situation in *Kurtomathrips*. By using different collection methods such as beating, sweeping and examination of plant hosts or by using Berlese funnels for leaf or ground litter, different proportions in the wing forms and additional species may be found.

See Stannard (1957, 1968) for a key to some genera and families of Nearctic thrips. Mound and Marullo (1996) provide an excellent discussion and keys to Neotropical thrips. The taxonomic arrangement used here follows this work. A total of 1450 slide mounts of newly collected specimens were studied. The slides are deposited in the collections of the US Department of Agriculture, Beltsville, MD.

Suborder Terebrantia
Family Merothripidae
The Large-legged Thrips
Genus *Merothrips* Hood 1912

M. morgani Hood 1912: 132.
> Distribution. Adventive?; eastern United States, Puerto Rico, Bermuda, South America, Panama, Hawaii, Australia, India, Kenya; Isabela, Santa Cruz.
> Bionomics. A fungivore, usually on twigs and branches; arid, transition and humid forest zones, in frass under bark; March–May.

Family Aeolothripidae
The Broad-winged or Banded Thrips
Genus *Franklinothrips* Back 1912

F. vespiformis (Crawford) 1909: 109 (*Aeolothrips*)
> Distribution. Indigenous; southern USA to Brazil; Isabela, Pinzón, Santa Cruz.
> Bionomics. A predator on thrips and mites; arid to pampa zone, and agriculture zone; March, May–June, August.

Family Thripidae
The Common Thrips
Genus *Baileyothrips* Kono and O'Neill 1964

B. limbatus (Hood) 1925: 124 (*Anaphothrips*).
  Distribution. Indigenous; Florida, Hawaii, Guatemala to Panama, Trinidad; Marchena, Pinta.
  Bionomics. Phytophagous on *Chamaesyce* species, *Euphorbea* species; arid zone; in open grassy *Bursera* forest; March.

Genus *Caliothrips* Daniel 1904

C. phaseoli Hood 1912: 113 (*Heliothrips*)
  Distribution. Indigenous; southwestern USA, Cuba, Mexico to Argentina; Fernandina, Floreana, Isabela, Pinta, Santa Cruz.
  Bionomics. Phytophagous on plants in various families; littoral scrub on sand; arid, transition and evergreen shrub zones; also taken by net on boat from air between the islands (Peck 1994*a*); March–May.

C. insularis (Hood) 1928: 234.
  Distribution. Indigenous?; Antilles, Bermuda, Costa Rica to Brazil, Mauritius, USA (Florida); Santa Cruz.
  Bionomics. Phytophagous on various grass family plants; arid zone forest; April and May.

C. punctipennis (Hood) 1912: 135.
  Distribution. Indigenous; Mexico, USA (Florida, Texas, Hawaii); Isabela.
  Bionomics. Phytophagous on grasses; in brackish littoral meadow and pampa; March.

Genus *Chaetanaphothrips* Prieser 1925

C. orchidii (Moulton) 1907: 52 (*Euthrips*).
  Distribution. Indigenous?; Central and North America, Caribbean Islands, Ecuador, Surinam, greenhouses in Europe, Orient, Australia; Isabela.
  Bionomics. Phytophagous on various plants, damages bananas in Central America, on orchids; agricultural zone and humid forest; March and May.

Genus *Frankliniella* Karny 1910

F. rodeos Moulton 1933: 115; Bailey 1967: 203. The South American Rose thrips
  Distribution. Indigenous?; Argentina, Brazil; Santa Cruz, Pinzón.
  Bionomics. Flower-inhabiting; arid and transition zones; beating on vegetation and on *Cordia lutea* and *Scalesia affinis* leaves; January–February. Not collected in the present study. See Nakahara (1997).

F. schultzei (Trybom) 1910: 151 (*Physopus*).
   Distribution. Indigenous; Cuba, Jamaica, Puerto Rico, Trinidad, widely distributed in tropical and subtropical areas of the world; Santa Cruz.
   Bionomics. Polyphagous; agricultural zone, arid zone; April. See Nakahara (1997).

F. species undetermined (pale yellow)
   Distribution. Indigenous?; Española, Fernandina, Floreana, Isabela, Pinta, Rábida, Santa Cruz, Santiago.
   Bionomics. Phytophagous; littoral-arid zone and transition zone, to mossy humid forest; in *Cordia lutea*; February–May.

F. species (brown)
   Distribution. Indigenous?; Santiago, Santa Cruz.
   Bionomics. Polyphagous; In lower transition zone forest, humid forest, and mossy forest; May and June.

Genus *Heliothrips* Haliday 1836

H. haemorrhoidalis (Bouché) 1833: 42 (Thrips), the greenhouse thrips.
   Distribution. Indigenous; pantropical, subtropical, in temperate zone greenhouses; Isabela, Pinzón, San Cristóbal, Santa Cruz, Santiago.
   Bionomics. Polyphagous; in arid and agricultural, transition, evergreen shrub, humid forest, and pampas; sifting leaf and moss litter; January–June.

Genus *Neohydatothrips* John 1929

N. williamsi (Hood) 1928: 230 (*Sericothrips*)
   Distribution. Indigenous?; described from St. Croix, Virgin Islands; Floreana, Isabela, Santa Cruz.
   Bionomics. Polyphagous; upper arid zone, agricultural zone, humid forest; March–May.

N. species
   Distribution. Indigenous?; Floreana, Isabela, Marchena, Pinta, Santa Cruz.
   Bionomics. Arid to pampa zones; March–May.

Genus *Plesiothrips* Hood 1915

P. brunneus Hood 1936: 252.
   Distribution. Indigenous?; described from Panama; Isabela, Santa Cruz, Santiago.
   Bionomics. Known from grasses. Galápagos habitat: lower transition zone forest, and mossy humid zone forest; May and June.

P. perplexus (Beach) 1896: 216 (*Sericothrips*).
   Distribution. Indigenous; known from many countries in the Western Hemisphere; Isabela, Santa Cruz, Santiago.

Bionomics. Reported from various plants in grass family; Galápagos habitat: humid forest and shrub/pampa zones; April and May.

Genus *Scirtothrips* Shull 1909

S. panamensis Hood 1935: 153.
    Distribution. Indigenous; described from Panama and also recorded from Costa Rica; Fernandina, Marchena, Santa Cruz, Santiago.
    Bionomics. Phytophagous on miscellaneous vegetation; members of this genus feed mainly on foliage and shoots; also flowers and on fruitlets; mossy humid forest; March–May.

S. species (short PA setae)
    Distribution. Indigenous?; Isabela (Cerro Azul).
    Bionomics. Transition and pampa zones; May.

Genus *Scolothrips* Hinds 1902

S. pallidus (Beach) 1896: 226 (*Thrips*); Priesner 1950: 43–45; Bailey 1967: 204.
    Distribution. Indigenous?; North America; Santa Cruz.
    Bionomics. A predator on spider mites; in arid zone forest and on *Ipomoea* in transition zone; February.

S. sexmaculatus (Pergande) 1890: 539 (*Thrips*).
    Distribution. Indigenous?; North America and Jamaica; Pinzón.
    Bionomics. A predator on spider mites, in litter in pampa zone; March.

S. species
    Distribution. Indigenous?; Floreana, Isabela, Santa Cruz.
    Bionomics. Apparently a predator on spider mites; arid to agricultural zone, pampa zone; most of the specimens of *Scolothrips* collected in the survey were this species; March–June.

Genus *Selenothrips* Karny 1911

S. rubrocinctus (Giard) 1901: 264 (*Physopus*). The red-banded cacao thrips.
    Distribution. Indigenous; pantropical and subtropical; Floreana, Isabela, Santa Cruz, Santiago.
    Bionomics. Polyphagous; arid zone to pampa zone, and agricultural zone; also taken by net on boat in air between the islands (Peck 1994*a*), and this suggests it may be a indigenous species; January–June.

Genus *Thrips* Linnaeus 1758

T. species (1 male)
    Distribution. Adventive?; distribution?; Isabela.
    Bionomics. Polyphagous; agricultural zone, in carrion trap; June.

Suborder Tubulifera
Family Phlaeothripidae
Genus *Adraneothrips* Hood 1925

A. tibialis (Hood) 1914: 39 (*Haplothrips*).
Distribution. Adventive?; USA (Florida), Costa Rica, Cuba, Puerto Rico, Trinidad, Brazil, Peru; Santa Cruz.
Bionomics. Phytophagous (on grasses and leaves of sugarcane); arid to *Miconia* zone; February–June.

A. species, pale species
Distribution. Indigenous?; Rábida, Santa Cruz.
Bionomics. Phytophagous; arid to humid forest zone; May.

Genus *Allothrips* Hood 1908

A. megacephalus mexicanus Stannard 1955: 154.
Distribution. Indigenous?; Mexico and Costa Rica; Santa Cruz, Santa Fé.
Bionomics. Scavenger?; in littoral and arid zone *Bursera* forest under *Cryptocarpus*; March and May.

Genus *Antillothrips* Stannard 1957

A. cingulatus (Hood) 1919: 80 (*Zygothrips*).
Distribution. Indigenous; USA (Florida, Hawaii), Jamaica, Panama, Puerto Rico, Trinidad, Australia, Thailand; Floreana, Santa Cruz.
Bionomics. Phytophagous; occurs in deep grass; brackish littoral meadow, and humid forest zone; March.

Genus *Carathrips* Hood 1950

C. rufescens (Hood) 1942: 617 (*Hoplothrips*).
Distribution. Indigenous?; described from Panama; Fernandina, Floreana, Isabela, Marchena, Santa Cruz.
Bionomics. Phytophagous; type specimens collected on dead leaves and branches; arid and transition zone forests (flight intercept); March–May.

Genus *Gastrothrips* Hood 1912

G. harti (Hood) 1913: 162 (*Zygothrips*).
Distribution. Indigenous?; USA (Texas); Española, Santa Cruz.
Bionomics. Collected on dead branches; feeds on fungal spores; arid zone forest; April.

G. species near *callipus* Hood 1935: 182.
Distribution. Indigenous?; Española, Pinta, Santa Cruz.
Bionomics. Feeds on fungal spores; littoral and arid zone forests; April.

G. species
Distribution. Indigenous?; Santa Cruz.
Bionomics. Antennal segment III and occasionally IV of Galápagos specimens are partially yellowish but completely brown in types of *callipus*; arid zone; May.

Genus *Haplothrips* Amyot and Serville 1843

H. gowdeyi Franklin 1908: 724 (*Anthothips*); Bailey 1967: 205. The black flower thrips.
Distribution. Adventive? from Africa; pantropical and subtropical; Fernandina, Floreana, Genovesa, Isabela, Pinzón, Rábida, Santa Cruz.
Bionomics. Polyphagous and infesting flowers; arid to humid zone; on vegetation and in *Tournefortia* flowers; March–May.

H. sonorensis Stannard 1956: 25; Bailey 1967: 205.
Distribution. Indigenous; south western North America; Santa Cruz.
Bionomics. Phytophagous?; no host plant data; one apterous male on littoral zone rotting driftwood; this Bailey record was not recollected in this study.

H. species, Bailey 1967: 205.
Distribution. Indigenous?; Santa Cruz.
Bionomics. Phytophagous; arid zone; on vegetation; January. This record of Bailey was not confirmed, but several species of *Karnyothrips* were collected, and this was once treated as a subgenus of *Haplothrips* (Stannard 1968: 429).

Genus *Hoplandrothrips* Hood 1912

H. species
Distribution. Indigenous?; Isabela, Pinta, Santa Cruz.
Bionomics. Detritivore; shrub forest; transition zone forest (*Bursera*, *Trema*); upper transition forest; species in this genus are found on dead branches; January–May.

Genus *Hoplothrips* Amyot and Serville 1843

H. angusticeps (Hood) 1908: 367 (*Trichothrips*; Bailey 1967: 206 (*Phalaeothrips*); Stannard 1968: 455.
Distribution. Indigenous; central and eastern North America; Fernandina, Floreana, Isabela, San Cristóbal, Santa Cruz, Santiago.
Bionomics. Fungus feeder?; sometimes under tree bark on dead branches, and polypore fungi; littoral scrub on sand, transition zone to mossy humid forest zone; February–May.

Genus *Karnyothrips* Watson 1924

K. caliginosus Hood 1938: 231.
  Distribution. Indigenous?; described from Peru; Isabela, Santiago.
  Bionomics. Predator?; Guava and humid forest; some members of this genus
    are predaceous on scale insects; March and April.

K. melaleucus (Bagnall) 1911: 61.
  Distribution. Indigenous; widespread in the tropics; Fernandina, Floreana,
    Genovesa, Isabela, Marchena, Pinta, San Cristóbal, Santa Cruz.
  Bionomics. Predator on scales?; littoral scrub on sand, in fern litter, *Ceiba*
    litter, transition zone and agricultural zone; February–May.

K. near. *longiceps* (Hood)
  Distribution. Adventive?; Santa Cruz, Santiago.
  Bionomics. Predator on scales?; *Miconia* and pampa zone; February and
    April.

Genus *Leptothrips* Hood 1909

L. species, possibly *yaqui* Johansen
  Distribution. Indigenous?; Española, Fernandina, Pinta, Santa Cruz.
  Bionomics. Predator?; littoral to pampa zone; April, May.

Genus *Liothrips* Uzel 1895

L. species
  Distribution. Indigenous?; San Cristóbal.
  Bionomics. Arid zone.

Genus *Macrophthalmothrips* Karny 1922

M. helenae Hood 1934: 79.
  Distribution. Indigenous; Panama, Mexico, Cuba, USA (Florida, South
    Carolina, Texas); Isabela, San Cristóbal, Santa Cruz.
  Bionomics. Fungivore?; arid and transition zones; January–May.

Genus *Nesothrips* Kirkaldy 1907

N. lativentris (Karny) 1913: 129.
  Distribution. Indigenous; widely distributed in the tropics; Fernandina,
    Floreana, Isabela, Pinta, San Cristóbal, Santa Cruz, Santiago.
  Bionomics. Feeds on fungus spores, littoral scrub on sand, litter under
    manchineel tree, arid zone, transition forest, humid forest; January–
    June.

Genus *Scopaeothrips* Hood 1912

S. unicolour Hood 1912: 70.
Distribution. Indigenous?; Honduras, Mexico, sw USA; Isabela.
Bionomics. Phytophagous on cactus?; arid forest; March.

Genus *Strepterothrips* Hood 1933

S. floridanus Hood 1938: 394 (*Arcyothrips*); Stannard 1957: 58; Bailey 1967: 207.
Distribution. Adventive?; Guyana, Trinidad, Mexico, Cayman Islands, Cuba, USA (Florida, Texas; Mound and Ward 1971); Española, Floreana, Isabela, Marchena, San Cristóbal.
Bionomics. Fungivore on freshly dead wood; littoral zone, upper arid zone, transition zone, and in *Ceiba* litter; February–April.

S. species
Distribution. Indigenous?; Española, Isabela, Marchena, Santiago, Santa Cruz.
Bionomics. Fungivore; arid zone *Bursera* forest; open *Cordia* woodland; March–May.

Genus *Trachythrips* Hood 1929

T. astutus Cott 1956: 100.
Amphibolothrips astutus Bailey 1967: 207; *Trachythrips astutus* Mound 1972a: 101).
Distribution. Indigenous?; known from USA (California); South Plaza.
Bionomics. Fungivore?; arid zone; known from leaf mold; collected in litter?; Bailey's record was not examined. Members of this genus live in leaf litter and apparently feed on fungal hyphae or break-down products associated with fungal infestation (Mound 1972a: 84).

T. species near *watsoni* Hood
Distribution. Indigenous?; Isabela.
Bionomics. Fungivore?; fern-moss litter; pampa zone. The posterior margin of the mesonotum of Galápagos specimens are straighter and mid and hind legs are darker than on *watsoni*.

Genus *Trichinothrips* Bagall 1929

T. species
Distribution. Indigenous?; Pinta, Santa Cruz.
Bionomics. Members of this genus are known from dead branches or grass; arid to transition zone, and *Miconia* zone; March–June.

Genus *Williamsiella* Hood 1925

W. morgani (Hood) 1941: 206 (*Phthirothrips*); this genus keys to *Phthirothrips* in Stannard (1957, 1968)

Distribution. Indigenous?; Costa Rica, Panama, Mexico, Florida; Isabela, Pinzón.

Bionomics. Many species in this genus are associated with mosses; in litter; March, May, and June.

Acknowledgements

The identifications of the new material are by Sueo Nakahara (Systematic Entomology Lab, US Department of Agriculture, Beltsville, MD). Bruce Heming, University of Alberta, Edmonton, graciously reviewed this chapter.

References

Bailey, S.F. 1967. A collection of Thysanoptera from the Galápagos Islands. Pan-Pac. Entomol. **43**: 203–210.

Kudo, I. 1985. The Japanese species of the genus *Chaetanaphothrips* Priesner (Thysanoptera, Thripidae). Kontyu **53**: 311–328.

Mound, L.A. 1972a. Species complexes and the generic classification of leaf litter thrips of the tribe Urothripini (Phlaeothripae). Aust. J. Zool. **20**: 83–103.

Mound, L.A. 1972b. Tropical fungus-feeding Thysanoptera of the genus *Macrophthalmothrips*. J. Entomol. (B) **41**: 77–88.

Mound, L.A. 1972c. Polytypic species of spore-feeding Thysanoptera in the genus *Allothrips* Hood (Phlaeothripidae). J. Aust. Entomol. Soc. **11**: 23–36.

Mound, L.A. 1989. Systematics of thrips (Insects: Thysanoptera) associated with mosses. Zool. J. Linn. Soc. **96**: 1–17.

Mound, L.A., and Houston, K.J. 1987. An anotated check-list of Thysanoptera from Australia. Occas. Pap. Syst. Entomol. **4**: 1–28.

Mound, L.A., and Marullo, R. 1996. The thrips of Central and South America: and introduction (Insecta: Thysonoptera). Mem. Entomol., Internal. **6**: 488 pp.

Mound, L.A., and O'Neill, K. 1974. Taxonomy of the Merothripidae with ecological and phylogenetic considerations (Thysanoptera). J. Nat. Hist. **8**: 481–509.

Mound, L.A., and Ward, A. 1971. The genus *Strepterothrips* Hood and its relatives with description of *S. tuberculatus* (Girault) comb.n. (Thysanoptera). J. Aust. Entomol. Soc. **10**: 98–104.

Nakahara, S. 1991. Two new species of *Caliothrips* (Thysanoptera: Thripidae) and a key to the Nearctic species. J. N.Y. Entomol. Soc. **99**: 97–103.

Nakahara, S. 1997. Annotated list of the *Frankliniella* species of the world (Thysanoptera: Thripidae). Contrib. Entomol., Int. **2**: 355–390.

Nakahara, S., and Hilburn, D.J. 1989. Annotated checklist of the Thysanoptera of Bermuda. J. N.Y. Entomol. Soc. **97**: 251–260.

Peck, S.B. 1994*a*. Aerial dispersal of insects between and to islands in the Galápagos Archipelago. Ann. Entomol. Soc. Am. **87**: 218–224.

Peck, S.B. 1994*b*. Sea surface (pleuston) transport of insects between the islands in the Galápagos Archipelago, Ecuador. Ann. Entomol. Soc. Am. **87**: 576–582.

Pitkin, B.R. 1977. The genus *Antillothrips* Stannard, with descriptions of two new species (Thysanoptera: Phlaeothripidae). Syst. Entomol. **2**: 53–58.

Priesner, H. 1950. Studies on the genus *Scholithrips* (Thysanoptera). Bull. Soc. Fouad. 1[er] Entomol. **34** 39–68, 7 figs.

Stannard, L.J. 1952. Phylogenetic studies of *Franklinothrips* (Thysanoptera; Aeolothripidae). J. Wash. Acad. Sci. **42**: 14–23.

Stannard, L.J. 1956. Five new thrips from the southwest (Thysanoptera: Tubulifera). Proc. Biol. Soc. Wash. **69**: 21–28.

Stannard, L.J. 1957. The phylogeny and classification of the North American genera of the suborder Tubulifera (Thysanoptera). Ill. Biol. Monogr. **25**: 1–200, 144 Figures.

Stannard, L.J. 1968. The thrips or Thysanoptera of Illinois. Illinois Nat. Hist. Surv. Bull. **29**: 215–2552, 310 Figures.

Chapter 19
Order Homoptera

The Leafhoppers, Planthoppers, Aphids, and Scale Insects

17 families, 70 genera, 141 species
(71 endemic?, 13 indigenous?, 57 adventive?)
Minimally 50? natural colonizations

The Galápagos are now known to have a large and diverse Homoptera fauna, but much of it is adventive and was probably carried to the islands on plants by human settlers. The leading publications on Galápagos Homoptera are Walker (1851), Stål (1859), Butler (1877), Morrison (1924), Osborn (1924), Van Duzee (1933, 1937), Fennah (1967) and Williams (1977). The Cicadellidae classification used here follows the work of Freytag (in prep.). The classification of Coccidae follows Williams (1977). Entomopathogenic fungi on Galápagos insects, mostly on Coccidae, are reviewed by Evans and Samson (1982). General data on Coccidae is in Ben-Dov (1993) and on other scale insects in Williams and Watson (1988*a*, 1988*b*, 1990) and Williams and Granara de Willink (1992). Because all female scale insects are wingless and are poor natural dispersers, I assume that most scale insect species in the Galápagos were carried there on plants by humans. The only other recent paper I know of on the scale insect fauna of an oceanic island is that for Bermuda (Hodgson and Hilburn 1991). It is also of special note that aphids were not previously reported from the Galápagos. Several genera of leafhoppers and plant hoppers seem to have experienced substantial speciation in the Galápagos; more so than in any other group of insects. Very little information is available on the host associations of all of these plant feeding insects. Within the smaller orders of Galápagos insects, the least studied group is presently the Homoptera. Large collections of Delphacidae, Issidae, Cixiidae, Cicadellidae and scale insects are still under study. Until these studies are completed it will be difficult to more accurately estimate the number of natural colonizations, and if the several species swarms are monophyletic.

Suborder Auchenorrhyncha
Superfamily Cicadoidea
Family Cicadellidae
The Leafhoppers

All leafhoppers, as immatures and adults, are phytophagous. They suck juices and cell contents from vessels and leaves and soft tissues of a wide variety of plants, and some may be host specific. Some may be serious agricultural pests, and can transmit plant diseases. It is not known if any Galápagos species are host specific. The collections were mostly made by malaise and flight and black light traps, and by sweeping and beating of vegetation. The new records are based on over 12 000 specimens identified by Paul H. Freytag, who is continuing study of the collections.

Subfamily Agalliinae
Genus *Agallia* Curtis 1833

A. new species A.
   Distribution. Adventive; known from mainland Ecuador; Fernandina, Floreana, Isabela, Santa Cruz, Santiago.
   Bionomics. Transition to pampa and inversion zones; January–October.

A. new species B.
   Distribution. Adventive; known from mainland Ecuador; San Cristóbal.
   Bionomics. Not known.

A. *plana* (Butler) 1877: 91 (*Jassus*).
   Distribution. Endemic; Floreana, Santa Cruz.
   Bionomics. Not known.

A. *striolaris* (Butler) 1877: 91 (*Jassus*).
   Distribution. Endemic; Floreana, Santa Cruz.
   Bionomics. Not known.

Genus *Austroagallia* Evans 1935

In addition to the following species, there are two undescribed species, endemic to either Fernandina or Isabella.

A. *mera* (Van Duzee) 1937: 123 (*Agallia*); Linsley and Usinger 1966: 138 (*Agalliopsis*); Viraktamath 1972: 1.
   A. *arrhenonigra* Viraktamath 1972: 2.
   Distribution. Endemic; Santa Cruz, Santiago.
   Bionomics. Humid forest zone; March–May.

Subfamily Coelidiinae
Tribe Teruliini
Genus *Coelidiana* Oman 1937

C. *krameri* Freytag 2000: 325
   Distribution. Adventive; Floreana, Isabela, Santa Cruz.
   Bionomics. Transition and humid forest zones; February–June; malaise and flight traps. Unknown from mainland Ecuador but probably introduced from there.

Genus *Docalidia* Nielson 1979

D. spangleri Nielson 1979: 292.
  Distribution. Endemic; Floreana, Isabela, San Cristóbal, Santa Cruz.
  Bionomics. Arid to pampa zone; 100–1000 m; malaise, FIT, yellow pan
  traps, and sweeping; January–May.

Genus *Jikradia* Nielson 1979

J. galapagoensis (Osborn) 1924: 77; Nielson 1979: 88.
  Jassus galapagoensis Osborn 1924: 77; Linsley and Usinger 1966: 138
  (*Coelidia*).
  Jassus infestus Van Duzee 1937: 124
  Distribution. Adventive?; Mexico to Panama; Isabela, Santa Cruz, Santiago.
  Bionomics. Littoral and arid zones; in grass-sedges at side of coastal salt la-
  goon, and on and under mangrove trees; sweeping, pan traps, and at
  lights; January–May.

Subfamily Deltocephalinae
Tribe Scaphytopiini
Genus *Scaphytopius* Ball 1931

S. aequinoctialis (Van Duzee) 1933: 30 (*Platymetopius*).
  Distribution. Endemic; Fernandina, Floreana, Genovesa, Isabela, Marchena,
  Pinta, Pinzón, Santiago, Santa Cruz, San Cristóbal.
  Bionomics. Arid to pampa and agriculture zones; March–July. Probably on
  trees or shrubs.

S. obliquus (Walker) 1851: 851 (*Acrocephalus*).
  S. retusus (Van Duzee) 1937: 123 (*Platymetopius*).
  Distribution. Endemic; Fernandina, Isabela, Pinta, San Cristóbal, Santa
  Cruz, Santiago.
  Bionomics. Transition to evergreen shrub zones, but most abundant in hu-
  mid forest zone; January–June. Probably on trees or shrubs.

S. new species
  A series of 15 additional undescribed endemic species each occur singly or
on a combination of the islands Darwin, Fernandina, Floreana, Genovesa,
Isabela Marchena, Pinta, Rábida, San Cristóbal, Santa Cruz., and Santiago.
They probably all feed on trees or shrubs.

Tribe Macrostelini
Genus *Cicadulina* China 1926

C. tortilla Caldwell 1952: 78; Abedrabbo et al. 1990: 122.
  Distribution. Adventive; widespread throughout the Americas; Fernandina,

Floreana, Isabela (Sierra Negra), San Cristóbal, Santa Cruz, Santiago, Seymour.

Bionomics. Host for the dryinid wasp *Gonotopus fernandinae* Olmi; arid to pampa zones; January–December. A species of economic importance.

Genus *Macrosteles* Fieber 1866

M. fascifrons (Stål) 1858: 194 (*Jassus*). New archipelago record.
   Distribution. Adventive?; North, Central, and South America; Isabela, Santa Cruz, San Cristóbal, Santiago.
   Bionomics. Not known.
   Note. This genus is being revised by Kwon.

Tribe Balcluthini
Genus *Balclutha* Kirkaldy 1900

B. aridula Linnavuori 1959: 344. New archipelago record.
   Distribution. Adventive; Central and western South America; Fernandina, Isabela, Rábida, Santiago.
   Bionomics. Not known.

B. incisa (Matsumura) 1902: 360 (*Gnathodus*). New archipelago record.
   Distribution. Adventive; worldwide; Española, Fernandina, Isabela, Rábida, Santa Cruz, Santiago.
   Bionomics. Littoral to humid forest and agriculture zones; February–May; malaise traps and at kerosene lamps.

B. lucida (Butler) 1877: 91 (*Jassus*). New archipelago record.
   B. floridana (DeLong and Davison) 1933: 66 (*Eugnathodus*)
   Distribution. Adventive; worldwide; Fernandina, Isabela, Marchena, Pinta, Rábida, San Cristóbal, Santa Cruz, Santiago.
   Bionomics. Arid to pampa and agriculture zones; February–August; very abundant and widespread; parasitized by Dryinidae wasps and the strepsipteran *Elenchus koebeli*.

B. neglecta (DeLong and Davidson) 1933: 55 (*Eugnathodus*). New archipelago record.
   Distribution. Adventive; North, Central, South America, Philippines, Fernandina, Isabela, Pinzón.
   Bionomics. Arid to pampa and agriculture zones; February–May.

B. rosea (Scott) 1876: 83 (*Gnathodus*). New archipelago record.
   Distribution. Adventive; worldwide; Española, Floreana, Isabela, Marchena, Pinta, San Cristóbal, Santa Cruz, Santiago.
   Bionomics. Arid to pampa and agriculture zones; February–May; parasitized by Dryinidae wasps and strepsiptera (*Elenchus koebeli*).

Balclutha-like new species 1 and 2. New archipelago records.
Distribution. Endemic; Isabela.; and Fernandina, Isabela, Santa Cruz.
Bionomics. Not known.

Tribe Deltocephalini
Genus *Amplicephalus* DeLong 1926

A. insularis (Van Duzee) 1933: 30 (*Deltocephalus*)
Distribution. Endemic; Fernandina, Floreana, San Cristóbal, Santa Cruz.
Bionomics. Unknown.

A. new species 1 and 2. New archipelago records.
Distribution. Endemic?; one on San Cristóbal, the other on Fernandina.
Bionomics. Not known

Genus *Circulifer* Zachvatkin 1935

C. tenellus (Baker) 1895: 24 (*Thamnotettix*). New archipelago record.
Distribution. Adventive; widespread in New World and Mediterranean areas;
Bartolomé, Caamaño, Española, Floreana, Genovesa, Isabela, Marchena,
Pinta, Pinzón, Rábida, San Cristóbal, Santa Cruz, Santa Fé, Seymour.
Bionomics. From 0 to 40 m elevation; in littoral, mangrove, and arid zones
and village habitats; by UV light, FIT, pan, and malaise traps;
February–June.

Genus *Exitianus* Ball 1929

E. fasciolatus (Melichar) 1911: 107 (*Athysanus*).
E. digressus (Van Duzee) 1933: 32 (*Athysanus*).
Distribution. Adventive; South America; Fernandina, Floreana, Isabela,
Rábida, San Cristóbal, Santa Cruz, Santiago, Seymour.
Bionomics. From 0 to 1320 m; coastal lagoon grass margins, villages, agri-
culture areas, through all vegetation zones to pampa-grassland and in-
version zones; by UV light, malaise, sweeping, and FIT traps; March–
June; parasitized by dryinid wasps.

Genus *Sanctanus* Ball 1932

S. discalis (Van Duzee) 1933: 31 (*Scaphoideus*).
Distribution. Endemic; Fernandina, Floreana, Marchena, Pinta, Santiago.
Bionomics. From 0 to 400 m elevation; in littoral, arid, and transition zone hab-
itats; by FIT, malaise and pan traps, sweeping and light traps; March–May.
Notes. Placed in *Ampliacephalus* by Blocker et al. 1995, Ann. Entomol. Soc.
Am. 88: 294.

Subfamily Neocoelidiinae
Genus *Coelidiana* Oman 1938

C. new species.
   Distribution. Endemic?; Isabela, Santa Cruz.
   Bionomics. From littoral and arid zones, through agricultural regions, to pampa zone; by FIT, malaise, pan traps, sweeping, and uv lights; January–June.
   Notes. The species is probably adventive, but there are no mainland records.

Subfamily Typhlocybinae
Genus *Empoasca* Walsh 1862

E. canavalia DeLong 1932: 114. New archipelago record.
   Distribution. Adventive; North, Central and South America; Fernandina, Floreana, Genovesa, Isabela, Marchena, Pinta, Pinzón, San Cristóbal, Santa Cruz, Santiago.
   Bionomics. Along coastal areas.

Subfamily Xestocephalinae
Genus *Xestocephalus* Van Duzee 1892

X. desertorum (Berg) 1879: 262 (*Athysanus*). New archipelago record.
   Distribution. Adventive; North, Central and South America; Fernandina, Floreana, Isabela, Marchena, Pinta, San Cristóbal, Santa Cruz, Santiago, Seymour.
   Bionomics. Widespread and common in arid through pampa zones; January–December.

Superfamily Fulgoroidea
The Planthoppers

These phytophagous insects, represented by five families in the Galápagos (of the 18 families worldwide), are one of the more important groups of endemic species. Two of these families of fulgoroids also reached the Hawaiian Islands, where they also radiated and became one of the more important components of that fauna (Asche 1997). The Hawaiian species are usually oligophagous or monophagous on woody dicots, but nothing is known of their hosts in the Galápagos.

Family Cixiidae

Key to genera of Galápagos Cixiidae (by H. Hoch)

1a Mesonotum tricarinate; body compressed; tegmina steeply tectiform; female ovipositor long, orthopteroid-like, tergite IX without distinct wax bearing area . . . . *Nymphocixia*

1b Mesonotum pentecarinate; body depressed; tegmina shallowly tectiform; female ovipositor short, tergite IX truncate, with distinct wax-bearing area . . . . *Oliarus*

Genus *Nymphocixia* Van Duzee 1923

N. vanduzeei galapagensis Fennah 1967: 58.
  N. unipunctata of authors, (not Van Duzee 1933).
  Distribution. Indigenous; South America; endemic subspecies; Fernandina, Isabela, Santa Cruz.
  Bionomics. Arid zone; January–March.

Genus *Oliarus* Stål 1862

This is a virtually worldwide genus with many species. Some 58 species have descended from a single colonization of Hawaii (Asche 1997). The patterns of the Galápagos species include one entry of subterranean habitats and loss of eyes. H. Hoch is studying the collections.

O. agrippa Fennah 1967: 62.
  Distribution. Endemic; Santa Cruz.
  Bionomics. Arid zone; January–February.

O. alabandus Fennah 1967: 61.
  Distribution. Endemic; San Cristóbal.
  Bionomics. Arid zone; February.

O. alastor Fennah 1967: 63.
  Distribution. Endemic; Floreana.
  Bionomics. Arid zone; February.

O. galapagensis Van Duzee 1933: 33; Fennah 1967: 58.
  Distribution. Endemic; Isabela.
  Bionomics. Arid zone; on *Scalesia gummifera*; January.

O. hernandesi Hoch and Izquierdo 1996: 1496.
Distribution. Endemic; Floreana.
Bionomics. Arid zone; in lava tube caves of Post Office Bay (Barn Owl and Finch Caves); eyeless.

O. remansor Fennah 1967: 60.
Distribution. Endemic; Pinzón (summit).
Bionomics. Transition zone; February.

O. new species.
Distribution. A series of an additional 6 undescribed endemic species occur on: Santa Cruz (arid zone, 20–140 m); Santa Cruz (arid zone, 20 m); Santiago (humid forest, 520–740 m); Pinzón (transition forest, 380 m); Fernandina (transition to humid forest, 110–1000 m); Española (arid zone, 5–20 m).

Family Issidae
Genus *Philatis* Stål 1862

This genus represents a remarable "species swarm," and all species have functionless or reduced flight wings. A preliminary study by M. R. Wilson indicates that another 5–18 new species are present, each a single island endemic, and each restricted to a single elevation-vegetation zone. This is then the largest species swarm and diversity of endemic species in the Galápagos biota. There is a need for additional study of the new material and its evolutionary origin and patterns.

P. athamas Fennah 1967: 92.
Distribution. Endemic; Darwin.
Bionomics. Arid zone; January.

P. atrax Fennah 1967: 24(A–G).
Distribution. Endemic; Santa Cruz.
Bionomics. Pampa zone; February, April; under *Jaegiria hirta*.

P. auson Fennah 1967: 100.
Distribution. Endemic; Pinzón.
Bionomics. On *Scalesia* species; February.

P. breviceps Van Duzee 1933: 33; Fennah 1967: 82.
Distribution. Endemic; Floreana.
Bionomics. Arid zone; April.

P. cinerea Osborn 1924: 78; Fennah 1967: 83.
Distribution. Endemic; Genovesa.
Bionomics. Arid zone; March.

P. crockeri (Van Duzee) 1937: 119 (*Euthiscia*); Fennah 1967: 85.
Distribution. Endemic; Santa Cruz.
Bionomics. Unknown

P. daunus Fennah 1967: 98.
   Distribution. Endemic; Santiago.
   Bionomics. June.

P. delia Fennah 1867: 90.
   Distribution. Endemic; Santa Cruz.
   Bionomics. Arid zone; January–February.

P. deucalion Fennah 1967: 87.
   Distribution. Endemic; Santa Cruz.
   Bionomics. Arid zone; January–February.

P. latobius Fennah 1967: 94.
   Distribution. Endemic; Wolf.
   Bionomics. Arid zone; February.

P. lento Fennah 1967: 86.
   Distribution. Endemic; Floreana (Wittmer Farm).
   Bionomics. Humid forest zone; February.

P. lycambes Fennah 1967: 95.
   Distribution. Endemic; Pinzón (summit and caldera).
   Bionomics. Transition zone; February.

P. major Osborn 1924: 79; Fennah 1967: 82.
   Distribution. Endemic; Santa Cruz.
   Bionomics. Arid zone; February.

P. monaeses Fennah 1967: 34, 96.
   Distribution. Endemic; San Cristóbal (Progresso).
   Bionomics. Humid forest zone?; February.

P. opheltes Fennah 1967: 97.
   P. productus Van Duzee (not Stal) 1933: 34 (pars.).
   Distribution. Endemic; Seymour.
   Bionomics. Arid zone; June.

P. productus (Stål) 1859: 278 (*Mycterodus*); Osborn 1924: 78; Fennah 1967: 81.
   Distribution. Panama, Peru; indigenous; Genovesa.
   Bionomics. Arid zone; April.

P. rostrifera (Butler) 1877: 90 (*Issus*); Distant 1909: 73 (*Galápagosana*); Fennah 1967: 85.
   Distribution. Endemic; Floreana.
   Bionomics. Unknown.

P. servus (Van Duzee) 1933: 34 (*Philates*); Fennah 1967: 85.
   Distribution. Endemic; Isabela.
   Bionomics.

P. varia (Walker) 1851: 372 (*Issus*); Fennah 1967: 84.
  Distribution. Endemic; San Salvador (= Santiago).
  Bionomics.

P. vicinus (Van Duzee) 1933: 34 (*Philates*); Fennah 1967: 84.
  Distribution. Endemic; Rábida.
  Bionomics. Arid zone.

Family Delphacidae

The following data present only published information. A key to genera is not given because generic limits of the named species need to be modified, and several new genera are present. The material is being studied by Manfred Asche.

Genus *Caenodelphax* Fennah 1965

C. teapae (Fowler) 1905: 135 (*Liburnia*); Fennah 1967: 77.
  Distribution. Indigenous; Mexico; Santa Cruz.
  Bionomics. Humid forest zone; no month data.

Genus *Nesosydne* Kirkaldy 1907

This genus seems to represent a remarkable "species swarm," and all species have functionless or reduced flight wings. There is a need for additional study of the new material and its evolutionary origin. There is also a swarm of 82 species in this genus in Hawaii (Asche 1997).

N. alcmaeon Fennah 1967: 69.
  Distribution. Endemic; Isabela, Santa Cruz.
  Bionomics. Arid to pampa zones; January–February. Host for *Elenchus* strepsiptera and dryinid wasp *Haplogonatopus crucianus* Olmi (Abedrabbo et al. 1990: 123.

N. augur Fennah 1967: 72.
  Distribution. Endemic; Fernandina (335 m).
  Bionomics. Humid forest; February.

N. brimo Fennah 1967: 65.
  Distribution. Endemic; Santa Cruz (Table Mt., 440 m).
  Bionomics. Humid forest; April.

N. iphis Fennah 1967: 68.
  Distribution. Endemic; Santiago (600 m), Santa Cruz.
  Bionomics. Humid forest; May. Host for dryinid wasp *Haplogonatopus crucianus* Olmi (Abedrabbo et al. 1990: 123).

N. olipor Fennah 1967: 73.
    Distribution. Endemic; Santa Cruz.
    Bionomics. Arid zone; February. Host for *Elenchus* strepsiptera and dryinid
        wasp *Haplogonatopus crucianus* Olmi (Abedrabbo et al. 1990: 123).

N. seneca Fennah 1967: 74.
    Distribution. Endemic; Fernandina (335 m).
    Bionomics. February.

N. simulans (Walker) 1851: 355 (*Delphax*); Fennah 1967: 67.
    Distribution. Endemic; Floreana, Santiago.
    Bionomics.

Genus *Peregrinus* Kirkaldy 1904

P. maidis (Ashmead) 1890: 323 (*Delphax*).
    Distribution. Adventive?; tropicopolitan, Santa Cruz.
    Bionomics. Humid forest zone; January–February.

Genus *Pissonotus* Van Duzee 1894

P. substituus (Walker) 1851: 354 (*Delphax*); Fennah 1967: 77.
    Delphax vicaria Walker 1851: 355
    Distribution. Endemic; Floreana, Santa Cruz.
    Bionomics. Humid forest zone; February.
    Note. Study of material by C. Bartlett (in litt. 12.VII.1996) shows that this
        species is not in the genus *Pissonotus*, but no other placement is sug-
        gested.

Genus *Sogatella* Fennah 1956

S. kolophon (Kirkaldy) 1907: 157 (*Delphax*); Fennah 1967: 76.
    Distribution. Adventive?; tropicopolitan; San Cristóbal, Santa Cruz.
    Bionomics. Humid forest zone; January–February.

Genus *Syndelphax* Fennah 1963

S. dissipatus (Muir) 1926: 33 (*Delphacodes*); Fennah 1967: 76.
    Distribution. Adventive?; tropical America; Santa Cruz.
    Bionomics. Arid to humid forest zone; January, April.

Family Tropiduchidae
Genus *Colgorma* Kirkaldy 1904

C. menalcas Fennah 1967: 79.
    Distribution. Endemic; Floreana (Wittmer Farm).
    Bionomics. Humid forest zone?; February.

Suborder Sternorrhyncha
Superfamily Psylloidea
Family Psyllidae
The Jumping Plantlice

This superfamily was not previously known from the islands. They all feed on the phloem tissues of their host plants. The identifications are courtesy of Dave Hollis, Natural History Museum, London, UK (NHMB).

Genus *Heteropsylla* Crawford 1914

H. cubana Crawford 1914: 46.
  Distribution. Adventive; Widespread, Central America, Caribbean, Surinam, USA (Florida), Hawaii, Pacific Islands, Australia, SE Asia, India, E Africa; San Cristobal.
  Bionomics. Hostplants: Mimosoid legumes – *Leucaena leucocephala, L. diversifolia, L. salvadorensis* (Muddiman et al. 1992); Galapagos host: *Cassia*.

Genus *Paracarsidara* Heslop-Harrison 1960

P. dugesii (Loew) 1886: 160 (*Carsidara*).
  Distribution. Adventive; Mexico, Guatemala, Cuba, Puerto Rico, Dominica, Jamaica, Trinidad; Santa Cruz.
  Bionomics. Recorded from *Wissadula* (Malvaceae) and an undetermined malvaceous host (Hollis 1987); Galapagos host: *Abutilon*.

Superfamily Aleyrodoidea
Family Aleyrodidae
The Whiteflies

This superfamily was not previously known from the islands. Identifications are courtesy of Andrew Jensen (SEL, BARC, USDA, Washington, DC).

Genus *Aleurothrixus* Quaintance and Baker 1914

A. floccosus (Maskell) 1895: 432. The woolly whitefly.
  Distribution. Adventive; Widely distributed in New and Old World tropics and subtropics; Isabela, Santa Cruz, Santiago.
  Bionomics. Arid to humid forest and agriculture zones, found on undersides of leaves; and in air over the sea; April and June. Hosts: reported on many plant species, and on numerous ornamental and horticultural plant species (Hamon 1981)

Genus *Bemisia* Quaintance and Baker 1914

B. spp.
    Distribution. Sombrero Chino
    Bionomics.

Superfamily Aphidoidea
Family Aphididae
The Aphids or Plantlice

In view of the large number of aphid species now known from the Galápagos, it is of interest that there are no earlier records of these insects. This is probably due to the need for careful searching of vegetation and alcohol preservation of specimens, and the difficulty of making identifications. All of the following are new archipelago records. Smith and Cermeli (1979) give a list of the aphids of the Caribbean, and Central and South America. The identifications here are courtesy of Eric Mau (ECORC, Agriculture and Agri-Food Canada, Ottawa).

Subfamily Aphidinae
Tribe Aphidini
Genus *Aphis* Linnaeus 1758

A. coreopsidis (Thomas) 1878: 7.
    Distribution. Adventive?; New World from Manitoba to Argentina; Fernandina, Isabela, Santa Cruz, Santiago.
    Bionomics. Plant sap; transition zone, arid zone, open *Cordia* woodland, palo santo forest, and *Scalesia* forest; collected by FIT, malaise, sweeping, UV light, and pan traps; April–June; host plants: elsewhere on *Bidens*, *Coreopsis*, *Eupatorium* and other composites.

A. craccivora Koch 1854: 124. The cowpea aphid or groundnut aphid.
    Distribution. Adventive; cosmopolitan; Fernandina, Isabela, Santa Cruz, Santiago.
    Bionomics. Plant sap; polyphagous (legumes preferred); arid zone, transition zone, sea pleuston (Santa Fé to Santa Cruz), Palo santo forest; FIT, malaise, pan, carrion and UV light traps; May–June. Facultatively anholocyclic (normally so in tropics).

A. gossypii Glover 1877: 36. The cotton aphid or cucumber aphid.
    Distribution. Adventive; cosmopolitan; Fernandina, Isabela, Marchena, Santa Cruz, Santiago.
    Bionomics. Plant sap; polyphagous; transition zone, humid forest, agricultural zone, arid zone, pampa, moss elfin forest, open *Cordia* woodland, palo santo forest; FIT, malaise, pan, carrion and yellow pan traps;

March–July. The most common aphid on Pacific islands. Pest of cotton, Curcurbitaceae, potatoes, among others. A complex of several lineages with different host preferences.

A. nerii Boyer de Fonscolombe 1841: 179. The oleander aphid.
   Distribution. Adventive; circumtropical/subtropical; Fernandina, Isabela, Pinta, Santa Cruz, Santiago.
   Bionomics. Plant sap; transition zone, arid zone, guyabillo groves, and *Scalesia* pasture; FIT; malaise and pan traps; March–June. Known from many Pacific islands. On Ascelpiadaceae and Apocynaceae, occasionally on other plants.

A. spiraecola Patch 1914: 253 (= *A. citricola* van der Groot 1912: 26). The green citrus aphid.
   Distribution. Adventive; cosmopolitan; Española, Fernandina, Isabela, Santa Cruz, Santiago.
   Bionomics. Plant sap; polyphagous (over 20 plant families worldwide); arid zone, pampa zone, transition zone, shrub forest, sparse grass, mossy forest and *Scalesia* pasture; collected with FIT, sweeping, malaise, carrion and pan traps; April–July. Primary hosts of holocyclic populations are *Citrus* and *Spiraea*; important pest of *Citrus*; a vector of 17 plant viruses. Both anoholocyclic and holocyclic populations exist.

Genus *Hysteroneura* Davis 1919

H. setariae (Thomas) 1878: 5. The rusty plum aphid.
   Distribution. Adventive?; originally North American, in recent decades becoming circumtropical/subtropical; Española, Fernandina, Genovesa, Isabela, Pinta, Rábida, Santa Cruz, Santiago.
   Bionomics. Plant sap; arid zone, palo santo forest, transition forest, sparse grass, littoral zone, agricultural zone, *Miconia* zone, *Scalesia* forest, and pampa zone; FIT, malaise, carrion and UV light traps; March–July. Found on grasses (usually at base of spike). Anholocyclic in tropics/subtropics; in North America there is host alternation, with *Prunus* spp. as primary host; secondary hosts are numerous species of Gramineae.

Genus *Rhopalosiphum* Koch 1854

R. maidis (Fitch) 1856: 550. The corn-leaf aphid.
   Distribution. Adventive; cosmopolitan; Isabela, Santa Cruz, Santiago.
   Bionomics. Plant sap; arid zone, transition zone, humid forest and pampa zone; FIT traps; April–June. Pest of aerial parts of grasses including crops, especially maize. Anholocyclic (except in apparent areas of origin in southwest foothills of Himalayas, where primary host is *Prunus* species).

R. rufiabdominale (Sasaki) 1899: 202.

  Distribution. Adventive?; circumtropical/subtropical; Isabela, Santa Cruz.

  Bionomics. Plant sap; upper transition and *Scalesia* forest; malaise and FIT; May. Found on roots of various grasses and sedges, also some Solanaceae and occasionally other herbaceous dicots; important pest of rice. Anholocyclic (except in some populations in Japan and probably other areas of southeast Asia, with alternation to *Prunus* spp.).

<h2 style="text-align:center">Tribe Macrosiphini
Genus Acyrthosiphon Mordvilko 1914</h2>

A. bidenticola C.F. Smith 1960: 157.

  Distribution. Adventive?; Puerto Rico, Brazil, Venezuela, Cuba, probably widespread in South America; Fernandina, Santa Cruz.

  Bionomics. Plant sap; transition and *Scalesia* zones; collecting by sweeping and with FIT and malaise traps; April–May.

<h2 style="text-align:center">Genus Aulacorthum Mordvilko 1914</h2>

A. circumflexum (Buckton) 1876: 130.

  Distribution. Adventive; circumtropical and common greenhouse pest elsewhere; Isabela (Cerro Azul).

  Bionomics. Plant sap; polyphagous; FIT; May. Originally oriental.

A. solani (Kaltenbach) 1843: 15.

  Distribution. Adventive; cosmopolitan (origin Europe); Santa Cruz.

  Bionomics. Plants sap; polyphagous; transition zone; FIT; May. Important pest of potato and tulips and of various plants in greenhouses.

<h2 style="text-align:center">Genus Dactynotus Rafinesque 1818</h2>

D. (*Lambersius*) species

  Distribution. Indigenous?; New World; Fernandina, Santa Cruz.

  Bionomics. Plant sap; *Scalesia*; collected by sweeping and with FIT-malaise; May.

  Note: Current catalogues, and consequently most authors, now use *Uroleucon* for *Dactynotus* on dubious grounds.

<h2 style="text-align:center">Genus Macrosiphum Passerini 1860</h2>

M. species

  Distribution. Indigenous?; Fernandina.

  Bionomics. Plant sap; *Scalesia* zone; May.

Genus *Sitobion* Mordvilko 1914

S. species
  Distribution. Endemic?; Santa Cruz.
  Bionomics. Plant sap; *Scalesia*; FIT-malaise; May–June. Matches none of the species previously recorded for South America; no indigenous South American species described.

Genus *Myzus* Passerini 1860

M. persicae (Sulzer) 1776: 105. The green peach aphid.
  Distribution. Adventive; cosmopolitan; Fernandina, Santa Cruz.
  Bionomics. Plant sap, polyphagous; transition zone, arid zone, *Scalesia*, Palo santo forest; FIT, malaise and pan traps; March–June. Principal aphid pest on many crops. Host alternating, in part, to peach and some other *Prunus* species where climate permits; otherwise anholocyclic.

Genus *Pentalonia* Coquerel 1859

P. nigronervosa Coquerel 1859: 260. The banana aphid.
  Distribution. Adventive; circumtropical/subtropical; Santa Cruz.
  Bionomics. Plant sap; *Scalesia*; FIT and malaise, May–June.

Family Hormaphididae
Subfamily Cerataphidinae
Genus *Cerataphis* Lichtenstein 1882

C. species
  Distribution. Adventive; Oriental in origin, widespread in tropics and in greenhouses; Santiago.
  Bionomics. Plant sap; widespread in tropics and in greenhouses (1 on orchids, 1 on palms); humid forest; FIT at Aguacate camp; June.

Family Pemphigidae
Subfamily Fordinae
Tribe Baizongini
Genus *Geopemphigus* Hille Ris Lambers 1933

G. ?floccosus (Moreira) 1925: 31.
  Distribution. Indigenous?; known from Caribbean and bordering continental areas, south to central Brazil; Fernandina, Isabela.
  Bionomics. Plant sap; FIT; May. Found on roots of *Ipomoea*. There are a few collections from composites which are reported to have slight mensural differences from the *Ipomoea* forms, and may belong to a second undescribed species.

Superfamily Coccoidea
The Scale Insects

This is a large group of small sized and highly modified insects which are attached to plant tissues. In general, scale insects have not been good at dispersing naturally to remote islands. Most of the Galápagos species have been introduced by humans on living host plants. Many can be serious economic pests. Additional data on some of the introduced and pest species can be found at the scalenet web site (http://www.sel.barc.usda.gov/scalenet/scalenet/htm). M.L. Wilson has my unstudied material.

Family Ortheziidae
The Ensign Coccids
Genus *Orthezia* Bosc d'Antic 1784

O. galapagoensis Kuwana 1902: 28; Morrison 1925: (redescription); Williams 1977: 89.
> Distribution. Endemic?; Duncan, Isabela, Santiago, Seymour.
> Bionomics. Herbivore; littoral and arid zones; April, December; Galápagos host plants: *Bursera graveolens*, *Cordia lutea*, *Scalesia microcephala*, *Heliotropum angiospermum*, *Cryptocarpus pyriformis*.

O. insignis Browne 1887: 169; Williams 1977: 89.
> Distribution. Adventive; cosmotropical; San Cristóbal.
> Bionomics. Herbivore; littoral and arid zones; December; Galápagos host plant: *Verbena litoralis*; elsewhere on many host plants.

O. praelonga Douglas 1891: 246; Williams 1977: 89.
> Distribution. Adventive; Jamaica, Trinidad, Guyana, Brazil; Fernandina.
> Bionomics. Herbivore; February; Galápagos host plant unknown (elsewhere in world on many species, *Capsicum*, *Croton*, *Citrus*, etc.

Family Margarodidae
The Giant Coccids and Ground Pearls
Genus *Icerya* Maskell 1878

I. purchasi Maskell 1878: 220. The cottony cushion scale.
> Distribution. Adventive; worldwide in *Citrus* growing areas; indigenous to Australia; Baltra, Floreana, Isabela, Marchena, Pinzon, San Cristóbal, Santa Cruz, Santiago, Seymour Norte.
> Bionomics. Herbivore; littoral, arid, and agriculture zones, March–May; found on many host plants; in Galápagos on many trees, e.g., *Citrus*, *Scaevola*, *Acacia*, *Parkinsonia*, *Rhizophora*. It is now known from 15 species of endemic plants, 23 species of native plants, and 6 species of

introduced plants (Roque and Causton 2000). The species was first found in Galápagos in 1982, was introduced on ornamental *Acacia* trees, and reached pest proportions in Puerto Ayora in 1995–96. The newly hatched nymphs are the dispersal stage and they can be aerially dispersed. The species is now known from 80 countries, and feeds on over 200 species of plants (Hale 1970). Males are rare; females are hermaphrodites, with both ovaries and testes, and are self-fertilizing. Control is by quarantine, and biocontrol with *Rhodalia* lady-bird beetles.

Genus *Margarodes* Guilding 1829

M. similis Morrison 1924: 143; Williams 1977: 90.
    Distribution. Endemic?; Baltra, Eden, Santa Cruz.
    Bionomics. Herbivore; arid zone; April; Galápagos host plants: the insects are subterranean and on the rootlets of *Bursera malacophylla* and *Maytenus octogona*.

Family Pseudococcidae
The Mealybugs

Among the families of Coccoidea, the Pseudococcidae and the Halimococcidae are generally the only families to have naturally colonized oceanic islands. The Hawaiian Islands naturally have only these two families. In Hawaii, more than ten endemic genera and 70 endemic species descended from some six ancestral colonizations (Beardsley 1997). In contrast, only a few species seem to be endemic to the Galápagos. All members of the family are sessile plant parasites, except for the fragile and short-lived (and taxonomically ignored) males. Aerial dispersal seems unlikely for adults of these insects and natural colonization on sea-transported plant materials seems most likely for the native species.

Genus *Ferrisia* Fullaway 1923

F. virgata (Cockerell) 1893: 254; Williams 1977: 90.
    Distribution. Adventive; cosmopolitan; Baltra, Isabela, Santiago, Santa Cruz.
    Bionomics. Herbivore; arid zone; January, December; Galápagos host plants: *Tribulis cistoides, Waltheria ovata, Bursera graveolens, Coldenia fusca, Hippomane mancinella, Laguncularia racemosa, Hibiscus tiliaceus*; on many plant species worldwide.

Genus *Geococcus* Green 1902

G. coffeae Green 1933: 54; the coffee root mealybug.
    Distribution. Adventive; cosmopolitan; Santa Cruz.

Bionomics. Herbivore; agriculture zone; February; Galápagos host plant *Musa* species; on many plant species worldwide.

Genus *Phenacoccus* Cockerell 1893

P. parvus Morrison 1924: 147; Williams 1977: 90.
Distribution. Adventive; Neotropics and Pacific, Africa, Asia, Australia, Florida; Genovesa.
Bionomics. Herbivore; arid zone; April; bush near shore. The species was originally described from the Galápagos. It is adventive or maybe native to Galápagos, and has spread out of the Neotropics and is now known to be nearly circumtropical (Williams and Watson 1988*b*: 159, Williams and Granara de Willink 1992). It occurs on a long list of host plants.

P. species, Williams 1977: 90.
Distribution. ?, indigenous?; Floreana, Santa Cruz.
Bionomics. Herbivore; arid zone; February, December; Galápagos host plants: *Lecocarpus pinnatifidius*, *Cordia* species, unknown host.

Genus *Planococcus* Ferris 1950

P. citri (Risso) 1813: 416; Williams 1977: 90.
Distribution. Adventive; cosmopolitan; San Cristóbal, Santa Cruz.
Bionomics. Herbivore; arid zone; November, December; Galápagos host plants: *Blechnum brownei*, *Psidium guajava*, *Ricinus communis*, *Jatropha curcas*; on many plant species worldwide.

Genus *Pseudococcus* Westwood 1840

P. galapagoensis Morrison 1924: 148; Williams 1977: 90; Williams and Granara de Willink 1992: 447; Gimpel and Miller 1996: 56; the Galápagos mealybug.
Distribution. Endemic; Eden, Santa Cruz.
Bionomics. Herbivore; arid zone; January, April; hosts unknown, from roots of a plant.

P. insularis Morrison 1924: 150; Gimpel and Miller 1996: 64: the island mealybug.
Distribution. Endemic; Baltra.
Bionomics. Herbivore; April; under stone near brackish pool; arid zone.

P. species (unidentifiable immatures, probably of above species).
Distribution. Bartholomew, Culpepper, Duncan, Fernandina, Isabela, Santiago, Santa Cruz, Seymour, South Plaza.
Bionomics. Herbivore; January, February, May, December; Galápagos host plants: *Chamaesyce amplexicaulis*, *Scalesia incisa*, *Jasminocereus* species, *Laguncularia racemosa*, *Vallesia glabra* variety *pubsecens*, *Alternanthera filifolia*, *Scutia pauciflora*, grass, *Maytenus octogona*, miscellaneous beating.

Genus *Rhizoecus* Künckel d'Herculais 1878

R. insularis Hambleton 1976: 30; Williams 1977: 91; Williams and Granara de Willink 1992: 537.

Distribution. Endemic; Santa Cruz.

Bionomics. Herbivore; littoral and arid zones; January; host plant: *Hippomane mancinella*.

R. latus (Hambleton) 1946: 30 (*Morrisonella*); 1976: 32; Williams 1977: 91.

Distribution. Indigenous; Colombia, Ecuador; Santa Cruz.

Bionomics. Herbivore; littoral and arid zones; January; Galápagos host plants: on roots of *Hippomane mancinella*.

Note. Hambleton (1976: 30) indicates that another species (apparently undescribed), exists but is known only from an immature female.

Family Eriococidae
Genus *Eriococcus* Targioni 1868

E. papillosus Morrison 1924: 145; Williams 1977: 91.

Distribution. Endemic?; Bartholomew, Isabela, Santiago, Santa Cruz, Seymour.

Bionomics. Herbivore; April, May, December; Galápagos host plants: *Chamaesyce amplexicaulis*, *Coldenia fusca*, *Coldenia nesiotica*, *Euphoribia equisetiformis*, *Jasminocereus* species, *Cryptocarpus pyriformis*, *Heliotropum angiospermum*, *Waltheria ovata*.

Family Coccidae
The Scale Insects
Genus *Ceroplastes* Gray 1818

C. cirripediformis Comstock 1881: 333; Williams 1977: 92.

Distribution. Adventive; cosmopolitan; Baltra, Bartholomew, Eden?, Santiago, Santa Cruz, South Plaza.

Bionomics. Herbivore; arid zones; January, December; Galápagos host plants: *Maytenus octogona*, *Coldenia fusca*, *Laguncularia racemosa*, *Cryptocarpus pyriformis*, *Rhizophora mangle*, *Tournefortia* species; on many plant species worldwide.

Genus *Coccus* Linnaeus 1758

C. hesperidium Linnaeus 1758: 455; Kuwana 1902: 30 (*Lecanium hesperidium* variety *pacificum*); Williams 1977: 92.

Distribution. Adventive; cosmopolitan; arid zone; Baltra, Isabela, Pinta, Seymour.

Bionomics. Herbivore; March; Galápagos host plants: *Psychotria rufipes*, *Gossypium barbadense* variety *darwini*, *Musa paradisiaca sapientum*; on many plant species worldwide.

C. viridis (Green) 1889: 248; Williams 1977: 93.
  Distribution. Adventive; cosmopolitan; Santa Cruz.
  Bionomics. Herbivore; February; agriculture zone; Galápagos host plant: *Psidium* species; on many plant species worldwide.

Genus *Parasaissetia* Takahashi 1955

P. nigra (Nieter) 1861: 9; Williams 1977: 93.
  Distribution. Adventive; cosmopolitan; Santa Cruz, South Plaza.
  Bionomics. Herbivore; arid zone; May, June, December; Galápagos host plants: *Hibiscus* species, *Maytenus octogona*; on many plant species worldwide.

Genus *Pulvinaria* Targioni Tozzetti 1867

P. urbicola Cockerell 1893: 255; Williams 1977: 93.
  Distribution. Adventive; cosmopolitan; Santa Cruz.
  Bionomics. Herbivore; from tree hole; Galápagos host plant unknown; on many plant species worldwide.

Genus *Saissetia* Deplanche 1859

S. coffeae (Walker) 1852: 1079; Williams 1977: 93.
  Distribution. Adventive; cosmopolitan; Isabela, San Cristóbal, Santa Cruz.
  Bionomics. Herbivore; littoral and arid zones; February, December; Galápagos host plants: *Chiococca alba*, *Psychotria rufipes*, *Polypodium squamatum*, *Ricinus communis*, *Cordia lutea*, *Rhizophora mangle*, *Scalesia pedunculata*, grass; on many plant species worldwide.

S. miranda (Cockerell and Parrott) 1899: 12; Williams 1977: 93.
  Distribution. Adventive; cosmopolitan; San Cristóbal.
  Bionomics. Herbivore; December; Galápagos host plant: *Sida* species; on many plant species worldwide.

S. neglecta DeLotto 1969: 419; Williams 1977: 93; the Caribbean black scale.
  Distribution. Adventive; New World and Pacific; Bartolomé, South Plaza.
  Bionomics. Herbivore; arid zone; December; Galápagos host plant: *Maytenus octogona*; on many plant species worldwide.

Family Asterolecaniidae
The Pit Scales
Genus *Asterolecanium* Targioni 1868

A. pustulans (Cockerell) 1892: 143; Williams 1977: 93.
  Distribution. Adventive; cosmopolitan; Isabela.
  Bionomics. Herbivore, arid zone; Galápagos host plant: *Tournefortia pubescens*; December.

A. puteanum Russell 1935: 93; Williams 1977: 93.
Distribution. Adventive; eastern USA; Isabela.
Bionomics. Herbivore; arid zone; December; Galápagos host plants: *Croton scouleri* variety *scouleri, Waltheria ovata.*

Family Diaspididae
The Armored Scales
Genus *Aspidiotus* Bouché 1833

A. destructor Signoret 1869: 120; Williams 1977: 94.
Distribution. Adventive; cosmopolitan; Santa Cruz.
Bionomics. Herbivore; arid zone; January, December; Galápagos host plants: *Phoenix dactylifera, Vallesia glabra,* grass.

Genus *Chortinaspis* Ferris 1938

C. species, undescribed Williams 1977: 94.
Distribution. Endemic; Santa Cruz, Seymour.
Bionomics. Herbivore; arid zone; January, December; Galápagos host plant: *Opuntia echios* variety *zacana.*

Genus *Hemiberlesia* Cockerell 1897

H. lataniae (Signoret) 1869: 124 (*Aspidiotus*); Williams 1977: 94.
Distribution. Adventive; cosmopolitan; Baltra, Isabela, Santiago, Santa Cruz, Pinta.
Bionomics. Herbivore; littoral and arid zones; December; Galápagos host plants: *Cryptocarpus pyriformis, Waltheria ovata, Acacia macrancantha, Bursera graveolens, Scalesia incisa.*

Genus *Howardia* Berlese and Leonardi 1896

H. biclavis (Comstock) 1883: 98; Williams 1977: 94.
Distribution. Adventive; tropicopolitan; Santiago, San Cristóbal, Santa Cruz.
Bionomics. Herbivore; arid zone; December; Galápagos host plants: *Acacia macracantha, Citrus sinensis, Waltheria ovata.*

Genus *Lepidosaphes* Shimer 1868

L. beckii (Newman) 1867: 217; Williams 1977: 94.
Distribution. Adventive; cosmopolitan; Santa Cruz.
Bionomics. Herbivore; agriculture zone; May; *Citrus* species

Genus *Melanaspis* Cockerell 1897

M. odontoglossi (Cockerell); Kuwana 1902: 32 (*M. smilacis*); Ferris 1941: 362 (*M. obtusa*); Williams 1977: 94.
    Distribution. Adventive; cosmopolitan; Isabela, Santiago, Santa Cruz.
    Bionomics. Herbivore; littoral, arid and transition zones; January, December; Galápagos host plants: *Chusquea* species, *Croton scouleri scouleri*, *Scalesia affinis*, *Waltheria ovata*, *Conocarpus erecta*, *Alternantheria filifolia*, *Cryptocarpus pyriformis*, *Maytenus octogona*.

Genus *Odonaspis* Leonardi 1897

O. species ("*saccharicola* complex"); Williams 1977: 95
    Distribution. Endemic?; Santa Cruz.
    Bionomics. Herbivore; littoral zone; January; host plant: a shore grass.

Genus *Parlatoria* Targioni-Tozzetti 1868

P. crotonis (Douglas) 1887: 242; Williams 1977: 94 (*Dactylaspis*).
    Distribution. Adventive; cosmopolitan; Santa Cruz.
    Bionomics. Herbivore; arid zone; December; Galápagos host plant: *Croton scouleri*.

Genus *Pinnaspis* Cockerell 1892

P. strachani (Cooley) 1899: 54; Williams 1977: 95.
    Distribution. Adventive; cosmopolitan; Bartholomew, Fernandina, Isabela, Santiago, Pinta, Santa Cruz, Seymour.
    Bionomics. Herbivore; littoral arid zones; December; Galápagos host plants: *Chamaesyce amplexicaulis*, *Cryptocarpus pyriformis*, *Scutia pauciflora*, *Euphorbia* species, *Gossypium barbadense* variety *darwinii*, *Polygala andersonii*, *P. sancti-georgil* variety *oblanceolata*, *Polygala galapagea*, *Scalesia affinis*, *Walteria ovata*, *Coldenia fusca*, *Conocarpus erecta*, *Cordia lutea*, *Hippomane mancinella*, *Vallesia glabra* variety *pubescens*, *Scalesia incisa*, *Croton* species, *Hibiscus tilaceus*, *Tournefortia* species, *Neptunia plena*, *Parkinsonia aculeata*, grass.

Genus *Pseudaulacaspis* MacGillivray 1921

P. major (Cockerell) 1894: 43; Williams 1977: 96.
    Distribution. Adventive; cosmopolitan; Santiago.
    Bionomics. Herbivore; arid zone; December; Galápagos host plants: *Cordia lutea*, *Hippomane mancinella*.

Genus *Selenaspidis* Cockerell 1897

S. articulatus (Morgan) 1889: 352; Williams 1977: 96.
  Distribution. Adventive; cosmopolitan; Santiago, San Cristóbal, Santa Cruz.
  Bionomics. Herbivore; arid zone; December; Galápagos host plants: *Vallesia glabra* variety *pubescens*, *"Acacia,"* *Castela galapageia*, *Citrus limetta*.

Genus *Velataspis* Ferris 1937

V. species (undescribed). Williams 1977: 96.
  Distribution. Endemic?; Santiago, South Plaza.
  Bionomics. Herbivore; arid zone; December; Galápagos host plant: *Maytenus octogona*.

Acknowledgements

I thank those who have helped with identifications and references: M. Ashe (Humboldt University, Berlin, Germany), Delphacidae; D. Burkhardt (Basel Museum, Basel, Switzerland), Psyllidae; P.M. Freytag (University of Kentucky, Lexington, KY), Cicadellidae; S. Hasek (Florida Department of Agriculture, Gainesville, FL), *Icerya purchasi*; H. Hoch (Humboldt University, Berlin, Germany), Cixiidae; D. Hollis (British Museum of Natural History), Psyllidae; A. Jensen (SEL, BARC, US Department of Agriculture), Aleurodidae; E. Mau (ECORC, Agriculture and Agri-Food Canada, Ottawa, Canada), Aphidoidea; M.L. Williams (Auburn University, Auburn, AL), Coccidoidea; and M. Wilson (National Museum of Wales, Cardiff, Wales), Issidae.

References

Abedrabbo, S., Kathirithamby, J., and Olmi, M. 1990. Contribution to the knowledge of the Elenchidae (Strepsiptera) and Dryinidae (Hymenoptera: Chrysidoidea) of the Galápagos Islands. Boll. Inst. Entomol. "Guido Grandi" Univ. Bologna **45**: 121–128.

Asche, M. 1997. A review of the systematics of Hawaiian planthoppers (Hemiptera: Fulgoroidea). Pac. Sci. **51**: 366–376.

Beardsley, J.W. 1997. Hawaiian Pseudococcidae (Hemiptera): a group that Perkins missed. Pac. Sci. **51**: 377–379.

Ben-Dov, Y. 1993. A systematic catalogue of the soft scale insects of the World. Flora and Fauna Handbook no. 9, Sandhill Crane Press, Gainesville, FL.

Butler, A.G. 1877. Account of the Zoological collection made during the visit of H.M.S. "Peterel" to the Galápagos Isalnds. X. Lepidoptera, Orthoptera, Hemiptera. Proc. Zool. Soc. London 1877: 86–91.

Champion, G.C. 1924. The insects of the Galápagos Islands. Entomol. Mon. Mag. **60**: 259–260.

Distant, W.L. 1909. VIII—Rhynchotal Notes—XLVIII. Ann. Mag. Nat. Hist. ser. 8, **4**: 73–87.

Evans, H.C., and Samson, R.A. 1982. Entomogenous fungi from the Galápagos Islands. Can. J. Bot. **60**: 2325–2333.

Fennah, R.G. 1954. The higher classification of the family Issidae (Homoptera: Fulgoroidea) with descriptions of new species. Trans. R. Entomol. Soc. London, **105**: 455–474.

Fennah, R.G. 1967. Fulgoroidea from the Galápagos Archipelago. Proc. Calif. Acad. Sci., ser. 4, **35**: 53–102, 30 Figures, 1 table.

Ferris, G.F. 1941. Atlas of scale insects of North America, vol S: III-362. Stanford Univ. Press, CA.

Freytag, P.M. 2000. A new species of *Coelidiana* (Homoptera: Cicadellidae) from the Galápagos Islands. Entomol. News **111**: 325–327.

Gimpel, W.F., and Miller, D.R. 1996. Systematic analysis of the mealybugs in the *Pseudoccus maritimus* complex (Homoptera: Pseudococcidae) Contr. Entomol., International **2**: 1–163.

Hale, D. 1970. Biology of *Icerya purchasi* and its natural enemies in Hawaii. Proc. Hawaii Entomol. Soc. **3**: 533–550.

Hambleton, E.J. 1976. A revision of the New World mealybugs of the genus *Rhizoecus* (Homoptera, Pseudococcidae). U.S. Dept. Agric. Tech. Bull. **1522**: 1–88.

Hamon, A.B. 1981. Woolly whitefly, *Aleurothrixus floccosus* (Maskell). Fla. Dep. Agric. Consum. Serv. Entomol. Circ. 232: 2 pp.

Hoch, H., and Izquierdo, I. 1996. A cavernicolous planthopper in the Galápagos Islands (Homoptera: Auchenorrhyncha: Cixiidae). J. Nat. Hist. **30**: 1495–1502.

Hodgson, C.J., and Hilburn, D.J. 1991. An annotated checklist of the Coccoidea of Bermuda. Fla. Entomol. **74**: 133–146.

Hollis, D. 1987. A review of the Malvales-feeding psyllid family Carsidaridae (Homoptera). Bull. Brit. Mus (Nat. Hist). Entomol. **56**: 87–127.

Kuwana, S.J. 1902. Coccidae from the Galápagos Islands. J. New York Entomol. Soc. **10**: 28–33, 2 pls.

Linnavuori, R. 1959. Revision of the Neotropical Deltocephalinae and some related subfamilies (Homoptera). Ann. Zool. Soc. Zool. Bot. Fennicae "Vanamo," **20**: 370 pp., 144 figs.

Linsley, E.G. 1977. Insects of the Galápagos (Supplement). Occas. Pap. Calif. Acad. Sci. **125**: 1–50.

Linsley, E.G., and Usinger, R.L. 1966. Insects of the Galápagos Islands. Proc. Cal. Acad. Sci. (4) **33**: 113–196.

Melichar, L. 1906. Monographie der Issiden (Homoptera). Abendlungen der Kaiserlichen-Koniglichen Zoologisch-Botanische Gesellschaft Wien **3** (4): 1–327, 75 figs.

Metcalf, Z.P. 1943. General Catalogue of the Homoptera. Fasc. IV, part 3, pp. 552. Agric. Res. Ser, U.S. Dep. Agric., Wash., DC.

Metcalf, Z.P. 1954. General Catalogue of the Homoptera. Fasc. IV, pt. 14, pp. vii + 54. Agric. Res. Ser, U.S. Dep. Agric., Wash., DC.

Metcalf, Z.P. 1958. General Catalogue of the Homoptera. Fasc. IV, part 15, vii 561. Agric. Res. Ser, U.S. Dep. Agric., Wash., DC.

Morrison, H. 1924. The Coccidae of the Williams Galápagos Expedition. Zoologica, **5**: 143–152, Figs. 33–37.

Morrison, H. 1925. Classification of scale insects of the subfamily Ortheziine. J. Agr. Res. **30**: 97–154.

Muddiman, S.B., Hodkinson, I.D., and Hollis, D. 1992. Legume-feeding psyllids of the genus *Heteropsylla* (Homoptera: Psylloidea). Bull. Entomol. Res. **82**: 73–117.

Muir, F. 1919. Notes on the Delphacidae in the British Museum Collection. Can. Entomol. **51**: 6–8.

Muir, F. 1919. On the Genus *Liburnia* White (Homoptera, Delphacidae). Proc. Hawaii Entomol. Soc. **4**: 48–50.

Neilson, M.W. 1979. A review of the subfamily Coelidiinae (Homoptera: Cicadellidae). III. Terulini, new tribe. Pac. Insects Monogr. **35**: 1–329.

Osborn, H. 1924. Homoptera of the Williams Galápagos Expedition. Zoologica **5**: 77–79.

Roque Albelo, L., and Causton, C. 2000. El Niño and introduced insects in the Galápagos Islands: different dispersal strategies, similar effects. Noticias de Galápagos. In press.

Smith, C.F., and Cermeli, M.M. 1979. An annotated list of the Aphididae (Homoptera) of the Caribean Islands, and South and Central America. North Carolina Agric. Research Serv., Tech. Bul. 259. 131 pp.

Stål, C. 1859. Hemiptera species novas descripsit. *In* Kongliga Svenska Fregatten Eugenies Resa Omkring Jorden, under befäl af C.A. Virgen, aren 1851–3. Zoologi 1, Insecta, pp. 219–298.

Van Duzee, E.P. 1933. The Templeton Crocker Expedition of the California Academy of Sciences, 1932. no. 4. Characters of twenty-four new species of Hemiptera from the Galápagos Islands and the coast and islands of Central America and Mexico. Proc. Calif. Acad. Sci. ser. 4, **21**: 25–40.

Van Duzee, E.P. 1937. Hemiptera of the Templeton Crocker Expedition to Polynesia in 1934–1935. Proc. Calif. Acad. Sci. ser. 4, **22**: 111–126.

Viraktamath, C.A. 1972. A new species of *Austroagallia* Evans from the Galápagos Islands (Homoptera: Cicadellidae). Occas. Pap. Calif. Acad. Sci. **101**: 1–4, 8 Figures.

Walker, F. 1851. List of the specimens of Homopterous insects in the collection of the British Museum. London. vol. 2, pp. 261–636; vol. 3, pp. 637–907.

Williams, M.L. 1977. Scale insects of the Galápagos islands. Virginia Polytech. Inst. State Univ. Res. Div. Bull. **127**: 85–98.

Williams, D.J., and Granara de Willink, M.C. 1992. Mealybugs of Central and South America. CAB International. Institute of Entomology, Wallingford, Oxon, England.

Williams, D.J., and Watson, G.W. 1988*a*. The scale insects of the tropical South Pacific Region. Part 1. The armored scales (Diaspididae). CAB International Institute of Entomology, Wallingford, Oxon, England. 289 pp.

Williams, D.J., and Watson, G.W. 1988*b*. The scale insects of the tropical South Pacific Region. Part 2. The mealybugs (Pseudococcidae). CAB International Institute of Entomology, Wallingford, Oxon, England. 260 pp.

Williams, D.J., and Watson, G.W. 1990. The scale insects of the tropical South Pacific Region. Part 3. The soft scales (Coccidae) and other families. CAB International Institute of Entomology, Wallingford, Oxon, England. 267 pp.

Chapter 20
Order Hemiptera (Heteroptera)

The True Bugs

20 families, 71 genera (5 endemic), 131 species
(78 endemic, 35 indigenous, 18 adventive)
Minimally 83 natural colonizations

Froeschner (1985) summarized the true bug fauna of the Galápagos and provided many excellent illustrations and keys to genera. The entire fauna, as then known, consisted of 20 families, 61 genera, and 109 species, and he estimated that this originated from a mimimum of 81 natural colonizations. It is of interest to note that mainland Ecuador was then known to have 507 recorded species of Hemiptera-Hepteroptera, and that only 13 species (2%) were shared with the Galápagos (Froeschner 1981). Thus, the vast area of the mainland of Ecuador is seemingly only 4.65 times richer in species than the Galápagos. In view of the larger size and ecological richness of the mainland, this actually shows how little is really known of the continental Ecuadorian fauna of Hemiptera. Almost all of the 22 new Galápagos records reported here of species not known to Froeschner are judged to be adventives or indigenous. Only a few new endemic species have been found by my teams.

Colonization of the Galápagos was most likely by flight of adults, or by rafting of adults and nymphs. The biologies of true bugs in the Galápagos are varied: they are feeders on plant juices from phloem vessels in roots, stems, leaves, fruits, buds and flowers, as well as seed predators and predators on arthropods. There is very little information on host plants of the Galápagos plant-feeding species. The feeding habits summarized here are more often hypotheses than based on direct evidence. Becker (1992) has shown that differences in dietary specializations (predator versus herbivore) in true bugs seems not to differentially effect their ability to colonize islands, in contrast to beetles. The trophic categories of Galápagos Hemiptera drawn from the summary of Froeschner (1985), as analyzed by Becker (1992) for selected families, were 20 predaceous and 71 herbivorous species. My reanalysis of my more complete data set, and using only the native species, changes these figures to 27 predaceous and 61 herbivorous species, but this seems not to challenge Becker's general conclusions about island Hemiptera.

The principle references to Galápagos Hemiptera-Heteroptera are by Stål (1859), Butler (1877), Heidemann (1901), Barber (1925, 1934), Van Duzee (1933, 1937), and Usinger and Ashlock (1966). Many other papers after 1966 deal with various families gathered by the 1964 GISP collecting program. Froeschner (1981) reviews the true bug fauna of Ecuador and gives keys to many of the genera of Hemiptera of northern South America. In that paper he repeats earlier Galápagos records, but gives no new data on Galápagos species. Froeschner (1985) included many new Galápagos records collected by my late colleague, Dr. Robert Silberglied. Most of my collections are from FITs, sweeping, and beating, so I do not list these methods for each species.

There is flux in the arrangement and higher classification of Hemiptera families. Because of this Froeschner (1981, 1985) and Henry and Froeschner (1988) used an alphabetical arrangement of families. I use here what seems to be the best available phylogenetic system for higher taxa (Schuh and Slater, 1995). Subfamily and tribe units follow Henry and Froeschner (1988) unless otherwise indicated. I list genera and species alphabetically. The following keys are from Froeschner (1985) and I have adapted them to include the taxa found more recently in the Galápagos. Henry and Froeschner (1988) give more data on the families, genera, or species that also occur in North America.

This is the largest order covered in this book, and it is sometimes combined with the above Homoptera into a single order, the Hemiptera, with the suborders Homoptera and Heteroptera (equivalent to the Hemiptera as used here). I think it is useful to give here an outline of the following higher classification that I have used for the Galápagos Hemiptera (after Schuh and Slater 1995). Following each family name I give my present estimates for the minimum number of natural colonizations in the family.

Suborder Gerromorpha

Superfamily Mesovelioidea: Mesoveliidae, 3.
Superfamily Gerroidea: Gerridae, 6; and Veliidae, 1.

Suborder Nepomorpha

Superfamily Corixoidea: Corixidae, 2.
Superfamily Notonectoidea: Pleidae, 1.

Suborder Leptopodomorpha

Superfamily Saldoidea: Saldidae, 3.

Suborder Cimicomorpha

Superfamily Reduvioidea: Reduviidae, 6
Superfamily Miroidea: Miridae, 26; Tingidae, 3.
Superfamily Naboidea: Nabidae, 2.
Superfamily Cimicoidea: Anthocoridae, 5; Cimicidae, 0.

Suborder Pentatomomorpha

Superfamily Pentatomoidea: Cydnidae, 1; and Pentatomidae, 9.
Superfamily Lygaeoidea: Berytidae, 1; and Lygaeidae, 9.
Superfamily Pyrrhocoroidea: Pyrrhocoridae, 0.
Superfamily Coreoidea: Coreidae, 0; Rhopalidae, 4; and Stenocephalidae, 1.

Key to Families of Galápagos Hemiptera (Heteroptera)
(from Froeschner 1985, see figures in that paper)

1a Antenna shorter than head, not projecting beyond margin of head . . . 2
1b Antenna longer than head, extending beyond margin of head 3
2a Labium broadly triangular, short, nonsegmented, with transverse sulca-
tions; fore tarsus a spoon- or scoop-shaped pala; hind legs oar-like;
head overlapping pronotum . . . . Corixidae
2b Labium cylindrical and obviously segmented, without transverse sulca-
tions; fore tarsus segmented and not spoon- or scoop-like; hind legs not
oar-like; head never overlapping pronotum . . . . Pleidae
3a Antennae with five distinct segments; head with a marginal carina extending anteriorly from eyes 4
3b Antennae with 4 segments; head without a marginal carina from eye . 5
4a Tibia with rows of stout spines . . . . Cydnidae
4b Tibia without spines, may have rows of fine setae. . . . . Pentatomidae
5a Last tarsal segment distinctly projecting beyond insertion of claws . . 19
5b Last tarsal segment not projecting beyond insertion of claws 6
6a Front wing with many small cells, dorsal surface appearing lace-like
. . . . Tingidae
6b Front wings (when present) without numerous cells, dorsal surface not
lace-like. 7
7a Prosternum with sharp-sided, distinct groove along midline; beak
short, more-or-less convexly curved away from venter of head, at rest
its tip placed in prosternal groove . . . . Reduviidae
7b Posternum without a groove along midline; beak not shaped as in
above half of couplet 8
8a Body extremely slender, about 7 times as long as pronotal width; antennae and legs linear, almost thread-like, apices of first antennal segment and of femur noticeably thickened . . . . Berytidae
8b Body and appendages stouter than above 9
9a Eye (in dorsal view) with a deep, abrupt notch on inner margin
. . . . Saldidae
9b Eye not so notched 10

10a Wings fully developed, corium with outer apical angle set off as a "cuneus" by a transverse of oblique suture 11

10b Wings absent or variously developed, never with a cuneus 12

11a Ocelli distinct; beak (in lateral view) arising from apex of head
. . . . Anthocoridae

11b Ocelli absent; beak arising from ventral surface of head . . . . Miridae

12a Juga prolonged and finger-like, surpassing and contiguous with apex of clypeus . . . . Stenocephalidae

12b Juga not surpassing apex of clypeus 13

13a Wings reduced to small, transverse, oval pads (not divided into areas) across base of abdomen; anterior margin of pronotum deeply concave, broadly U-shaped . . . . Cimicidae

13b Wings, when present (subdivided into areas) longer than broad (even when not reaching apex abdomen); anterior margin of pronotum not or only gently concave 14

14a Beak arising from apex of head, bucculae absent. . . . . Nabidae

14b Beak arising from ventral side of head between bucculae 15

15a First 2 segments of beak short, their combined length about half as long as head, third segment longer than head and prothorax combined; wings present or absent; forewing membrane apparently without veins
. . . . Mesoveliidae

15b Beak with first 2 segments elongate, their combined length surpassing base of head, third segment shorter than head; wings present; forewing membrane with veins 16

16a Wing membrane with not more than 5 longitudinal veins . . . Lygaeidae

16b Wing membrane with more than 5 longitudinal veins 17

17a Ocelli absent. . . . . Pyrrhocoridae

17b Ocelli present, distinct. 18

18a Side of thorax between middle and hind leg with a conspicuous rounded pore bordered by distinct elevations; Galápagos species with a long, outcurved spine on side of head between eye and antennal tubercle. . . . . Coreidae

18b Side of thorax without a visible pore; head without spine described in other half of couplet . . . . Rhopalidae

19a Middle coxa equidistant from front and back coxae; posterior femur less than half as long as length of body . . . . Veliidae

19b Middle coxa much closer to back coxa than to front coxa; posterior femur as long as or longer than length of body . . . . Gerridae

Suborder Gerromorpha
Superfamily Mesovelioidea
Family Mesoveliidae
The Marsh Treaders or Pondweed Bugs

Key to Galápagos genera of Mesoveliidae

1a   Adults with fully formed wings . . . . *Mesovelia*
1b   Adults without fully formed wings 2
2a   Mesonotum prolonged in middle, distinctly longer than pronotum; without pair of platelike elevated carinae on ventral surface of head
. . . . *Mesovelia*
2b   Mesonotum shorter than pronotum; with a pair of platelike elevated longitudinal carinae on ventral surface of head . . . . . . *Darwinivelia*

Subfamily Mesoveliinae
Genus *Darwinivelia* Anderson and Polhemus 1980

D. fosteri Anderson and Polhemus 1980: 375; Froeschner 1985: 23.
   Distribution. Endemic; Santa Cruz.
   Bionomics. Predator; intertidal zone?, also on surface of pools in deep lava crevices such as Grieta Iguana at CDRS; April; only micropterous specimens are known.
   Notes. This genus was once thought to be endemic to the Galápagos, but other species in the genus are now known from Colombia and Brazil (Polhemus and Manzano 1992). The type specimen was a prey item of a *Halobates* sea-strider.

Genus *Mesovelia* Mulsant and Rey 1852

Key to Galápagos species of *Mesovelia*

1a   Body plus wing length smaller, 2.0–2.2 mm; midline of mesonotum without yellowish line . . . . *M. amoena*
1b   Body plus wing length larger, 3.0–3.3 mm; midline of mesonotum with yellowish line . . . . *M. hambletoni*

M. amoena Uhler 1894: 217; Gerecke et al. 1994: 102.
   Distribution. Indigenous; eastern, central and southern North America to West Indies and South America; Floreana, Isabela, San Cristóbal, Santa Cruz, Santiago.

Bionomics. Littoral to pampa zones; on the surface of fresh and brackish ponds; predators or scavenger on small arthropods or crustaceans at the water surface; March–July, December.

M. hambletoni Drake and Harris 1946: 8; Gerecke et al. 1994: 102.
Distribution. Indigenous; Ecuador mainland; Isabela.
Bionomics. Predator on surface of littoral and arid zone brackish water ponds; May.

Superfamily Gerroidea
Family Gerridae
The Water Striders

Key to Galápagos genera of Gerridae

1a Abdomen compact and reduced, abdomen tip barely extending beyond apex of meso and meta trochanters; salt water and open ocean habitats; adults always apterous . . . . *Halobates*
1b Abdomen not compact and reduced, abdomen tip extending far behind apex of meso and meta trochanters; freshwater or inland brackish water habitats; adults apterous or winged . . . . *Limnogonus*

Tribe Halobatini
Genus *Halobates* Eschscholtz 1822

These wingless bugs skate on the surface of the sea, where they are predators or scavengers on material on the sea surface. See Cheng (1997) and Herring (1961) for more details of biology and distribution of the Pacific species.

Key to known or potential Galápagos species of *Halobates*
(from Herring 1961)

1a Apex of abdomen differentiated into a genital apparatus with a distinct angle on each side; males 2
1b Apex of abdomen regularly angular, without a differentiated section; females 6
2a Left subgenital styliform process much shorter than right one, at base abruptly bent out at right angle to body axis; length 4.5 mm 3
2b Left and right subgenital styliform processes subequal in length, neither bent at right angles to body axis 4
3a Right subgenital styliform process virtually straight; length 4.5 mm
. . . . *H. micans*

3b   Right syliform process strongly incurved; length 5 mm (not yet reported from the Galápagos) . . . . *H. splendens*

4a   Left styliform process from near base, slightly bent outward, right process straight, directed posteriad; length 4.4 mm . . . . . *H. sobrinus*

4b   Left and right styliform processes similarly developed, neither directed outward 5

5a   Styliform processes about as long as genital capsule, their apices projecting slightly behind outline of lateral angle of dorsal genital plate; length 3.6 mm (not yet reported from the Galápagos) . . . . *H. sericeus*

5b   Styliform processes about two-thirds as long as subgenital capsule, their apices not projecting; length 3.5 mm . . . . *H. robustus*

6a   Meso- and metanota with abundant short, suberect, black bristles (best seen from posterior view); length 4.2 mm . . . . *H. robustus*

6b   Meso- and metanota without bristles. 7

7a   Larger, length 5 mm; anterior tarsus with its two segments about equal in length (not yet reported from the Galápagos) . . . . . . *H. splendens*

7b   Smaller, length less than 4.5 mm; anterior tarsus with first segment distinctly shorter than second 8

8a   Antennal segment II very slightly longer than III, about half as long as IV; length 3.4 mm (not yet reported from the Galápagos. . . *H. sericeus*

8b   Antennal segment II one-third longer than III, about two-thirds as long as IV; length 4.0 mm 9

9a   Middle and posterior femora, when projecting posteriad, with posterior femur reaching only three-fourths of way to apex of middle femur; length 4.0 mm . . . . *H. micans*

9b   Middle and posterior femora, when arranged as above, with posterior femur reaching much more (eight-ninths) of way to apex of middle femur; length 4.0 mm. . . . . *H. sobrinus*

H. micans Eschscholtz 1822: 107; Barber 1934: 289, 1943: 80; Herring 1961: 227; Froeschner 1985: 17.

   Distribution. Indigenous; temperate-tropical circumglobal oceanic; near Santiago, near San Cristóbal.

   Bionomics. Ocean surface predator; March.

H. robustus Barber 1925: 253, 1934: 289; Heidemann 1901: 369 (species); Beebe 1924: 83, 432 (species nov.); Champion 1924: 260 (species nov.); Van Duzee 1937: 118; Usinger 1938: 84; Herring 1961: 229, 282; Froeschner 1985: 17.

   Distribution. Endemic; Fernandina, Floreana, Isabela, San Cristóbal, Santa Cruz.

   Bionomics. Ocean surface predator; January–May, October; for biology and behavior see Cheng and Maxfield 1980: 43; Foster and Treherne 1980,

1982; Birch, Cheng and Treherne 1979: 33; Treherne and Foster 1980. *H. robustus* is the most frequently found species around the Galápagos.

H. sericeus Escholtz 1822: 108; Froeschner 1985: 17.
Distribution. Indigenous; throughout the sub-tropical and tropical Pacific; between Santa Fé and Santa Cruz; previously known from east and west of the Galápagos, but not from the Galápagos themselves.
Bionomics. Ocean surface predator and scavenger; March.

H. sobrinus White 1883: 46; Herring 1961: 229, 251; Froeschner 1985: 17.
Distribution. Indigenous; east Pacific oceanic coast from west Mexico to western South America between 30°N and 5°S; near Galápagos.
Bionomics. Ocean surface predator.

H. splendens Witlaczil 1886: 178; Froeschner 1985: 17.
Distribution. Indigenous; off the west coast of Peru and Chile; between Santiago and Isabela, and Isabela and Fernandina; previously unknown from the Galápagos themselves.
Bionomics. Ocean surface predator and scavenger; March.

Tribe Gerrini
Genus *Limnogonus* Stål 1868

L. franciscanus (Stål) 1859: 265 (*Gerris*); Gerecke et al. 1994.
Distribution. Indigenous; Mexico to South America, and West Indies; Isabela.
Bionomics. Predator on the surface of fresh water; first discovered in Galápagos in 1991, seemingly a natural colonization; June.

Family Veliidae
The Broad-shouldered Water Striders
Tribe Microveliini
Genus *Microvelia* Westwood 1834

Key to Galápagos species of *Microvelia*

1a Abdomen broad; last segment not delimited from preceeding segment by constriction . . . . females
1b Abdomen narrow; last segment delimited from preceeding segment by constriction; males 2
2a Male genitalia paramere with straight sharply pointed projection (illustration in Bachmann 1978: 132) . . . . *M. ashlocki*
26. Male genitalia paramere with upcurved bluntly pointed projection (illustration in Bachmann 1978: 133). . . . . *M. isabelae*

M. ashlocki Polhemus 1968*b*: 129; Bachmann 1978: 131; Froeschner 1985: 57; Gerecke et al. 1995: 102.

Distribution. Endemic; Darwin, Isabela, San Cristóbal, Santa Cruz.

Bionomics. Predator; on surface of pools in *Miconia* zone; and waters of arid zone lava crevices; March–May, December. The two endemic species in this genus can only be separated by the shape of the male parameres. Both seemingly descended from a common colonizing ancestor.

M. isabelae Bachmann 1978: 131; Froeschner 1985: 55; Gerecke et al. 1995: 102.

Distribution. Endemic; Isabela, San Cristóbal.

Bionomics. Predator; arid zone; on surface of pond 600 m inland from coast; March–May.

Suborder Nepomorpha
Superfamily Corixoidea
Family Corixidae
The Water Boatmen
Subfamily Corixinae
Genus *Trichocorixa* Kirkaldy 1908

Key to Galápagos species *Trichocorixa*

1a Eye smaller, interocular distance about one-half width of head; length of pala (blade-like, modified, anterior tarsus) equal to length of posteroventral margin of eye; male fore pala with shorter peg row; male right clasper ornamented; length 3.0–3.2 mm . . . . *T. beebei*

1b Eye larger, interocular distance less than one-half width of head; length of pala two-thirds as long as posteroventral margin of eye; male fore pala with longer peg row; male right clasper not ornamented; length 2.8–5.2 mm . . . . *T. reticulata*

T. beebei Sailer 1948: 306; Froeschner 1985: 13; Gerecke et al. 1995: 102.

Distribution. Endemic; Genovesa (in Arcturus Lake in island center), Isabela (in Lake Darwin at Tagus Cove, Pto. Albermarle, Puerto Bravo).

Bionomics. Mostly phytophagous (on algae or diatoms); aquatic; February, May.

T. reticulata (Guerin-Meneville) 1857: 423 (*Corisa*); Sailer 1946: 617, 1948: 343; Nieser 1969: 153, 1970: 66, 1975: 217; Bachmann 1978: 134; Froeschner 1985: 13; Gerecke et al. 1995: 102.

Distribution. Indigenous; widespread Neotropical; Rábida, Santa Cruz, Santiago.

Bionomics. Mostly phytophagous (on algae or diatoms); aquatic; littoral, arid and transition zones; in salty lagoons and pools up to 600 m from coast; at blacklight in arid zone; February–July, October–November.

Notes. Several species in this genus are physiologically adapted to high concentrations of salt in saline coastal lagoons and pools. At present, the two Galápagos species seem to have mutually exclusive distributions (on separate islands). The report of *T. reticulata* on Isla Isabela (Gerecke et al. 1995: 132) is an error. If the two species are mutually exclusive is an interesting ecological question.

Superfamily Notonectoidea
Family Pleidae
The Minute Backswimmers
Genus *Paraplea* Esaki and China 1928

P. puella Barber 1923: 75; Bachmann 1978: 135; Gerecke et al. 1995: 103.
   Distribution. Indigenous; Mexico, Central America, West Indies to Trinidad, Gulf states of USA (Drake and Chapman 1953); Floreana, Isabela, San Cristóbal, Santa Cruz, Santiago.
   Bionomics. Predators on small crustaceans and other arthropods; diminutive (1.8 mm long); under surface in ponds or still waters; arid to humid forest zones, and agriculture zone; March–May.

Suborder Leptopodomorpha
Superfamily Saldoidea
Family Saldidae
The Shore Bugs

Key to Galápagos genera of Saldidae (excluding undescribed genus)

1a   Elytral membrane with five nearly equally elongate cells . . . *Pentacora*
1b   Elytral membrane with only four cells . . . . *Saldula*

Subfamily Chiloxanthinae
Genus *Pentacora* Reuter 1912

P. sphacelata (Uhler) 1877: 434 (*Salda*); Cobben 1965: 181; Polhemus 1968a: 21; Polhemus and Chapman 1979: 21; Froeschner 1985: 53.
   Salda rubromaculata Heidemann 1901: 368; Barber 1925: 253, 1934: 289 (*Pentacora*).
   Distribution. Indigenous; southern Europe, eastern and southern North America, West Indies, Mexico; Baltra, Isabela, Santa Cruz (2 km east of CDRS).

Bionomics. Predator on small invertebrates; inhabitant of shoreline of fresh or saline water and coastal lagoons; January–April.

Subfamily Saldinae
Genus *Saldula* Van Duzee 1914

S. galapagosana Polhemus 1968*a*: 22; Froeschner 1985: 53.
  Distribution. Endemic; Isabela, San Cristóbal, Santa Cruz, Santiago.
  Bionomics. Predator on small invertebrates; arid zone; inhabitant of geyser pool in crater (765 m) on Volcán Alcedo; in pampa at 1000 m, Sierra Negra; January–June, December.

Undescribed Genus: endemic genus

Undescribed species not placeable in presently defined subfamilies.
  Distribution. Endemic; Fernandina.
  Bionomics. Predator; littoral zone; February. Taken by spraying pyrethrum insecticide on usually submerged rocks at low tide at beach at Cabo Hammond. Because similar collecting has not found it elsewhere in the Galápagos, and Cabo Hammond is a west facing site, the affinities of the genus may be with the west Pacific, and not to the east on the Neotropical mainland. Otherwise, the genus can be expected to be found eventually in similar environments on the mainland west coast of South America.

Suborder Cimicomorpha
Superfamily Reduvioidea
Family Reduviidae
The Assassin Bugs

These predatory bugs are rather well represented in the Galápagos. Only a few seem to have been adventively transported by humans. The names used here follow Maldonado Capriles (1990) which also has distributional data. Wygodzinsky (1966) should be consulted for keys and illustrations.

Key to Galápagos genera of Reduviidae

1a Anterior coxae longer than length of head; anterior tibia at rest held against ventral surface of femur, latter spined ventrally; subfamily Emesinae 2
1b Anterior coxae not longer than head; anterior tibia at rest not held against femur, latter not spined ventrally. 5

2a Anterior femur ventrally with basalmost spine at least twice as long as any other spines; adult body large (15–20 mm); adults winged or wingless 3

2b Anterior femur ventrally with basalmost spine subequal in length to the other long spines; adult body small (3–10 mm); adults winged 4

3a Prothorax surface distinctly granular; female with well developed hind lobe of prothorax and fully developed wings; male without both pronotal hind lobe and wings. . . . . *Ghinallelia*

3b Prothorax surface smooth or finely granular; both sexes without pronotal hind lobe and wings . . . . *Barce*

4a Body uniform pale colour; anterior trochanter with one or more spines . . . . *Ploiaria*

4b Body darkly coloured, with white coloured areas or ridges; anterior trochanter without spines. . . . . *Empicoris*

5a Ocelli absent; subfamily Saicinae . . . . *Tagalis*

5b Ocelli present 6

6a Transverse pronotal constriction before middle of pronotum; subfamily Harpactorinae . . . . *Repipta*

6b Transverse pronotal constriction behind middle of pronotum; subfamily Peiratinae. . . . . *Rasahus*

Subfamily Emesinae
The Thread-legged Bugs
Tribe Leistarchini
Genus *Ploiaria* Scopoli 1786

M. macrophthalma (Dohrn) 1860: 244 (*Luteva*); Villiers 1970: 229. Schaefer et al. 1980: 48; Froeschner 1985: 46.
Distribution. Indigenous?; tropicopolitan; Santa Cruz.
Bionomics. Predator on insects; arid and agricultural zones; at light; January–May, July, December.

Tribe Metapterini
Genus *Barce* Stål 1866

B. fraterna (Say) 1832: 33 (*Ploiaria*). New archipelago record.
Distribution. Adventive?; North America to Mexico, Cuba, Colombia, Ecuador (Wygodzinsky 1966: 443); Santa Cruz.
Bionomics. Predator on insects; in pit trap in roadside pasture at 480 m; April and May.

Genus *Ghinallelia* Wygodzinsky 1966

This genus has at least 9 species (some are undescribed) in the Galápagos. They seem to be descended from one ancestral colonization. All species are flightless and most seem to be restricted to a single island. The genus ranges from Florida and the West Indies to Brazil.

Key to described Galápagos species of *Ghinallelia*

1a Abdomen without projections; length 13 mm . . . . . . *G. galapagensis*
1b Abdomen with postero-lateral angle of each segment projecting as a distinct tooth 2
2a Posterior lobe of head with an elongate spine (length appearing equal to height of eye) on each side of midline 3
2b Posterior lobe of head with a low tubercle (length less than one-fourth height of eye) on each side of midline 5
3a Mesonotum widest at midlength, lateral margin (in dorsal view) convex for full length; length 10–11 mm . . . . *G. usingeri*
3b Mesonotum widest near basal fourth, lateral margin (in dorsal view) gently concave in anterior two thirds 4
4a Pronotum stout, its width half its length; length 10 mm . . . *G. schaeferi*
4b Pronotum slender, its width distinctly less than half (about 40%) of its length; length 10 mm . . . . *G. leleuporum*
5a Head (in dorsal view) with preocular margins straight, parallel; postocular margins nearly straight, gradually converging posteriorly from eyes; length 13 mm . . . . *G. wygodzinsky*
5b Head with preocular margins weakly (but noticeably) convex; postocular margins convexly converging from eyes; length 11 mm . . *G. vagvolgyianus*

G. galapagensis (Heidemann) 1901: 367; Barber 1934: 286 (both *Ghilianella*); Wygodzinsky 1966: 487; Villiers 1970: 231; Schaefer et al. 1980: 48; Froeschner 1985: 47
Distribution. Endemic; Española.
Bionomics. Wingless ground-living predator; arid zone; March–May.

G. leleuporum Villiers 1970: 231, 1978: 50; Schaefer et al. 1980: 48; Froeschner 1985: 47
Distribution. Endemic; Santa Fé.
Bionomics. Wingless ground-living predator; arid zone; March, October.

G. schaeferi Villiers 1978: 49; Schaefer et al. 1980: 49; Froeschner 1985: 48
Distribution. Endemic; Rábida.
Bionomics. Wingless ground-living predator; arid zone; April.

G. usingeri Villiers 1970: 235, 1978: 50; Froeschner 1985: 48
Distribution. Endemic; Floreana.
Bionomics. Wingless ground-living predator; humid forest zone; February.

G. vagvolgyianus Villiers 1978: 46; Schaefer et al. 1980: 48; Froeschner 1985: 48
Distribution. Endemic; Santiago, Rábida.
Bionomics. Wingless ground-living predator; arid and transition zones; March–June, December.

G. wygodzinsky Villiers 1970: 233; Froeschner 1985: 48
Distribution. Endemic; Pinta.
Bionomics. Wingless ground-living predator; arid to humid forest zones; March–May.

G. species A. Villiers 1970: 231; Heidemann 1901: 367; Barber 1934: 286; Schaefer et al. 1980: 48 (last 3 as *G. galapagensis*); Froeschner 1985: 48
Distribution. Endemic; Isabela.
Bionomics. Wingless ground-living predator; arid zone; adults are needed; March.

G. species B. Villiers 1970: 235; Froeschner 1985: 48
Distribution. Endemic; Pinzón.
Bionomics. Wingless ground-living predator; arid zone; adults are needed; August.

G. species C.
Distribution. Endemic; Santa Cruz.
Bionomics. Wingless ground-living predators; transition to pampa zones; in pit traps; January–December.

Tribe Ploiariolini
Genus *Empicoris* Wolff 1811

Key to Galápagos species of *Empicoris*

1a Posterior half of dark pronotum with two pronounced white longitudinal ridges; scutellum without spine . . . . *E. barberi*
1b Posterior half of molted-light brown pronotum without distinct fine white ridges; scutellum with spine. 2
2a Posterior half of pronotum with short, broad, low, white mark
. . . . *E. rubromaculatus*
2b Posterior half of pronotum with long, thin distinct white ridge *E. armatus*

E. armatus (Champion) 1898: 165 (*Ploiarodes*). New archipelago record.
Distribution. Indigenous?; reported from Florida, Jamaica, Puerto Rico, Guatemala and Panama (Wygodzinsky 1966: 370); Floreana.

Bionomics. Predator on small arthropods; in FIT in interior of *Scalesia* forest, 325 m; April.

E. barberi (McAtee and Malloch) 1923: 7 (*Ploiarodes*). New archipelago record.
Distribution. Indigenous; Puerto Rico, Cuba, Peru (Wygodzinsky 1966: 370); Marchena, Pinta, Santa Cruz.
Bionomics. Predator on small arthropods; arid to pampa zones; March–May, July; in flight intercept and malaise traps.

E. rubromaculatus (Blackburn) 1889: 349 (*Ploiarodes*). New archipelago record.
Distribution. Indigenous?; virtually circumtropical (Wygodzinsky 1966: 384); Floreana.
Bionomics. Predator on small arthropods; in FIT in interior of *Scalesia* forest, 325 m; April.

Subfamily Harpactorinae
Genus *Repipta* Stål 1859

R. annulipes Barber 1925: 250; Heidemann 1901: 366 (nymph, as *Cosmoclopius (Harpactor)*); Barber 1925: 250, 1934: 287; Van Duzee 1937: 115; Schaefer et al. 1980: 49; Froeschner 1985: 48
Distribution. Endemic; Fernandina, Isabela, Santa Cruz.
Bionomics. Predator; arid to humid forest zone; March, February–June.

Subfamily Peiratinae
Genus *Rasahus* Amyot and Serville 1843

R. hamatus (Fabricius) 1781: 381 (*Reduvius*). New archipelago record.
Distribution. Adventive?; USA (FL-TX), Mexico to Argentina, West Indies; Española, Florena, Isabela, San Cristóbal, Santa Cruz.
Bionomics. Predator on insects; I consider this to be a recent introduction because it is difficult to believe that this large widespread species was previously present, but missed by all previous entomologists; littoral to humid forest and agriculture zones; February–June.

Subfamily Saicinae
Genus *Tagalis* Stål 1858

T. seminigra Champion 1898: 179, 180. New archipelago record
Distribution. Adventive; Central (and S.?) America; Santa Cruz.
Bionomics. Predator; agriculture zone, in guava thicket; July; FIT.

Superfamily Miroidea
Family Miridae
The Plant Bugs

This is the largest family of Heteroptera, and their habits and biologies are varied. The fauna of 43 species may have come from 26 ancestral colonizations. The classification used here is that of Schuh's (1995) catalogue. My unidentified material is being studied by M.D. Schwartz (ECORC, Agriculture and Agri-Food Canada, Ottawa), and he provided some of the following identifications.

Key to Galápagos genera of Miridae (adapted from Froeschner 1985)

1a Pronotum with a sharp impressed subapical groove from side to side, this groove setting off a distinct but narrow collar 2

1b Pronotum without such a groove or collar. 12

2a Head between eyes with impressed groove along midline 3

2b Head without impressed groove along midline 5

3a Pronotal collar and scutellum with noticeable, erect pubescence; beak very long, reaching beyond tip of abdomen . . . . *Galapagomiris*

3b Pronotal collar and scutellum without erect pubescence 4

4a Pronotum with posterolateral angles slightly but distinctly projecting, making lateral and posterior margins close to them appear concave; preocular part of head acutely triangular, projecting anterior to eyes by a distance greater than horizontal length of an eye . . . . *Fulvius*

4b Pronotum with posterolateral angles not projecting; preocular part of head obtusely rounded, projecting anterior to eyes by a length less than horizontal diameter of eye . . . . *Creontiades*

5a Pronotal disc, corium (except along costal margin) and scutellum polished, without setae or pubescence (except along costal margin). *Horcias*

5b Pronotal disc, corium, and often scutellum with one or two types of fine vestiture 6

6a Dorsal surface of corium with decumbent silky or wooly setae between suberect setae . . . . *Polymerus*

6b Dorsal surface of corium without decumbent setae between suberect setae. 7

7a Posterior tarsus with segment I longer than lengths of segments II and III combined 8

7b Posterior tarsus with segment I shorter than combined lengths of segments II and III 9

8a Suberect setae on antennal segment I at least half as long as diameter of segment I. . . . . *Dolichomiris*

8b Suberect setae on antennal segment I less than half as long as diameter
    of antennal segment I . . . . *Trigonotylus*
9a Pronotum (when not deflexed) posteriorly reaching to transverse groove
    on scutellum; eye in contact with anterior margin of pronotum 10
9b Pronotum posteriorly not extended to transverse groove on scutellum;
    eye at least narrowly separated from anterior margin of pronotum . . 11
10a Left end of male clasper with a small but strong, decurved hook
    . . . . *Taylorilygus*
10b Left end of male clasper without a hook. . . . . *Dagbertus*
11a Posterior margin of eye removed from pronotum by a space equal to or
    greater than diameter of antennal segment II; postocular part of head
    with a broad, longitudinal black line behind eye. . . . . . *Macrolophus*
11b Posterior margin of eye removed from pronotum by a space less than
    the diameter of antennal segment II; postocular part of head without a
    black line . . . . *Engytatus*
12a Head, pronotum, and scutellum mostly fuscous to black, strongly con-
    trasting with yellowish, subhyaline elytra; head with a large yellow
    spot next to inner margin of each eye . . . . *Tytthus*
12b Head, when black, without yellow spot next to each eye 13
13a In dorsal view, posterior margin of eye removed from pronotum by a
    space more than half the horizontal diameter of an eye. . *Galapagocoris*
13b In dorsal view, posterior margin of eye much closer to pronotum. . . 14
14a Pretarsus with lamellate, convergent parempodia . . . . . . *Sthenaridea*
14b Pretarsus with setiform, parallel parempodia 15
15a Corium translucent; dorsum with simple setae, but without flattened or
    scale-like setae . . . . *Campylomma*
15b Corium opaque; pronotum and corium bearing small, simple setae in-
    terspersed with flattened, more or less scale-like setae, the latter easily
    detached and lost (often must be searched for the few remaining ones)
    . . . . *Rhinacloa*

Subfamily Cylapinae
Tribe Cylapini
Genus *Fulvius* Stål 1862

Key to Galápagos species of *Fulvius*

1a Bases of corium and clavus, to level of scutellar apex, white, thence
    brown except for white spot on cuneal fracture; anterior coxa, except
    apex, brown; length 2.8–3.2 mm . . . . *F. brevicornis*

1b   Base of corium to level of scutellar apex brown, thence with a transverse white band; entire clavus and remainder of corium, except white
spot on cuneal fracture, brown; anterior coxa, except narrow base and
apex, white; length 2.9–3.2 mm . . . . *F. geniculatus*

F. brevicornis Reuter 1895: 138; Carvalho and Gagné 1968: 153; Froeschner
1985: 26.
   Distribution. Indigenous?; nearly tropicopolitan; Fernandina, Isabela, Santa
   Cruz, Santiago.
   Bionomics. A tree-trunk frequenting predator?, or fungus associate?; arid to
   pampa zones; January–July.

F. geniculatus Van Duzee 1933: 29; Barber 1934: 287; Carvalho and Gagné
1968: 152. Schaefer et al. 1980: 47; Froeschner 1985: 26.
   Distribution. Endemic; Baltra, Fernandina, Floreana, Isabela, Marchena,
   Pinta, Rábida, San Cristóbal, Santa Cruz, Santiago.
   Bionomics. Predator or fungus associate?; arid to pampa and agricultural
   zones; grass bases, and under *Jaegeria hirta*; January–December.

Subfamily Orthotylinae
Tribe Orothotylini
Genus *Galapagocoris* Carvalho 1968 (endemic genus)

G. crockeri (Van Duzee) 1933: 29; Barber 1934: 287 (both *Diaphnidia*); Carvalho
and Gagné 1968: 180; Froeschner 1985: 34.
   Distribution. Endemic; Fernandina, Floreana, Isabela, Marchena, Pinta, San
   Cristóbal, Santa Cruz, Santiago.
   Bionomics. Phytophagous?; arid to pampa and agricultural zone; January–June.

Subfamily Phylinae
Tribe Leucophoropterini
Genus *Tytthus* Fieber 1864

T. parviceps (Reuter) 1890: 259 (*Cyrtorhinus*); Froeschner 1985: 38.
   Distribution. Adventive?; Europe to South Africa; southern USA to Para-
   guay; Santa Cruz.
   Bionomics. Phytophagous; arid zone; July.

Tribe Phylini
Genus *Campylomma* Reuter 1878

C. citrina Carvalho, in Carvalho and Gagné 1968: 156; Froeschner 1985: 37
   Distribution. Endemic; Floreana, Santa Cruz.
   Bionomics. Phytophagous?; humid forest and agriculture zones; February.

Genus *Rhinacloa* Reuter 1876

Key to Galápagos Species of *Rhinacloa*

1a Male antennal segment II longer than width of head across both eyes;
female antennal segment II cylindrical, about as long as width of head
across both eyes . . . . *P. rubescens*

1b Male antennal segment II as long as or shorter than width of head
across both eyes; female antennal segment II widened toward apex,
length less than width of head across both eyes 2

2a Dorsum pale, medially with a broad brown to black stripe extending
back from apex of head and covering median area of pronotum, all of
scutellum and all of clavus; length 2.3–2.7 mm . . . . *R. usingeri*

2b Dorsum without a median stripe. 3

3a Beak very long, in male reaching to genital capsule, in female reaching
onto base of ovipositor; length 2.2–2.5 mm . . . . *R. longirostris*

3b Beak much shorter, not or only very slightly surpassing posterior coxae . 4

4a Entire dorsum pale yellowish, colour may appear darkened if most of
the easily abraded dark setae remain on the specimen; femora (except
spots) and entire coxae pale; length 2.3 mm . . . . *R. insularis*

4b Dorsum mostly black to reddish brown; femora, except extremities,
and all except apices of coxae black; length 2.3 mm . . . . . . *R. mella*

R. insularis (Barber) 1925: 250, 1934: 287; Carvalho and Gagné 1968: 163;
Schaefer et al. 1980: 47; Froeschner 1985: 37 (all preceeding as *Psallus*); Schuh
& Schwartz 1985: 419

    Distribution. Endemic; Darwin, Española, Fernandina, Floreana, Isabela,
      Pinta, San Cristóbal, Santa Cruz, Santiago.

    Bionomics. Phytophagous; host plants: *Alternanthera* species, *Scalesia* spe-
      cies; arid to pampa zone; at lights; January–May.

    Phylogeny. In a group with *R. longirostris, usingeri,* and *crassitoma* (of Ar-
      gentina, Brazil, and coastal Peru) (Schuh and Schwartz 1985). The spe-
      cies in this genus probably descended from two ancestral colonizations.

R. longirostris (Carvalho), in Carvalho and Gagné 1968: 158; Froeschner 1985:
38 (both as *Psallus*); Schuh and Schwartz 1985: 421

    Distribution. Endemic; Floreana, Isabela, Pinzón, San Cristóbal, Santa Cruz.

    Bionomics. Phytophagous; host plant: *Scalesia gummifera*; arid to transition
      zones; February, May–June.

R. mella (Van Duzee) 1937: 117 (*Europiella*); Carvalho and Gagné 1968: 159
(*Psallus*); Schuh and Schwartz 1985: 422

    Distribution. Endemic; Fernandina, Santa Cruz.

    Bionomics. Phytophagous; arid to pampa zones; February–April.

Phylogeny. In a group with *R. rubescens* and two species from the Florida Keys, and Mexico to Brazil (Schuh and Schwartz 1985).

R. rubescens (Carvalho), in Carvalho and Gagné 1968: 164; Froeschner 1985: 38 (all as *Psallus*; Schuh and Schwartz 1985: 427
   Distribution. Endemic; Floreana.
   Bionomics. Phytophagous; host plant: *Euphorbia*?; arid zone; February.

R. usingeri (Carvalho), in Carvalho and Gagné 1968: 162; Froeschner 1985: 38 (both as *Psallus*); Schuh and Schwarz 1985: 428
   Distribution. Endemic; Floreana, Santa Cruz.
   Bionomics. Phytophagous; humid forest zone; February–June.

Tribe Pilophorini
Genus *Sthenaridea* Reutter 1885

S. vulgaris (Distant) 1893: 448 (*Jornandes*); New archipelago record.
   Distribution. Adventive; Mexico, Guatemala, West Indies; Fernandina, Isabela, Santa Cruz, Santiago.
   Bionomics. Phytophagous; arid to pampa zone; sweeping; first records in 1991; May–July.

Subfamily Bryocorinae
Tribe Dicyphini
Genus *Engytatus* Rueter 1876

Key to Galápagos species of *Engytatus* (from Carvalho and Gagné 1968)

1a Colour pale flavescent to greenish, the extreme apices of corium and cuneus black or fuscous; segment II of antennae basally and median ring on segment I, black to fuscous. . . . . *E. modesta*
1b Colour yellowish to pale translucent; antennae unicolourous 2
2a Males 3
2b Females 7
3a Antenna segment I subequal to width of vertex; apex of dorsal bifurcation of pygophore short and blunt and with an apical tuft of setae
   . . . . *E. gummiferae*
3b Antenna segment I greater than width of vertex; apex of dorsal bifurcation of pygophore tapering and finger-like, variously clothed but never with an apical tuft of setae 4
4a Rostrum not surpassing apices of hind coxae. 5
4b Rostrum surpassing apices of hind coxae. 6

5a   Exceeding 2.60 mm in length; dorsal bifurcation of pygophore gla-
     brous with apex acute . . . . *E. affinis*
5b   Less than 2.60 mm long; dorsal bifurcation of pygophore with 3 distal
     setae dorsally, apex rounded . . . . *E. helleri*
6a   Antenna II longer than width of pronotum; lacking two distinct fields
     of bristles on dorsum of left clasper; dorsal bifurcation of pygophore
     strongly hooked distally and with two setae near apex . . . . . *E. arida*
6b   Antenna II subequal or less than width of pronotum; 2 distinct fields of
     bristles on dorsum of left clasper; dorsal bifurcation of pygophore
     gradually curved inwardly and not hooked downwardly at its apex,
     glabrous . . . . *E. floreanae*
7a   Antenna II subequal or less than width of vertex . . . . . *E. gummiferae*
7b   Antenna II greater than width of vertex 8
8a   Rostrum surpassing coxal bases, reaching base of abdominal segment
     five 9
8b   Rostrum not surpassing coxae, reaching posterior of metacoxae at the
     most 10
9a   Species less than 2.35 mm long . . . . *E. helleri*
9b   Species greater than 2.35 mm long . . . . *E. arida*
10a Antenna II usually less than width of pronotum at base; species less
     than 2.60 mm long . . . . *E. floreanae*
10b Antenna II usually greater than width of pronotum at base; species
     greater than 2.60 mm long . . . . *E. affinis*

E. affinis Gagné, in Carvalho and Gagné 1968: 176; Froeschner 1985: 35
(*Cyrtopeltis*).
     Distribution. Endemic; Santa Cruz.
     Bionomics. Phytophagous; host plant: *Scalesia affinis*; humid forest zone;
     January.

E. arida Gagné, in Carvalho and Gagné 1968: 176; Froeschner 1985: 35
(*Cyrtopeltis*).
     Distribution. Endemic; Santa Cruz.
     Bionomics. Phytophagous; humid forest zone; host plant: *Scalesia* species;
     January.

E. floreana Gagné, in Carvalho and Gagné 1968: 176; Froeschner 1985: 35
(*Cyrtopeltis*).
     Distribution. Endemic; Floreana.
     Bionomics. Phytophagous; humid forest zone; host plant: *Scalesia* species;
     February.

E. gummiferae Gagné, in Carvalho and Gagné 1968: 172; Froeschner 1985: 35 (*Cyrtopeltis*).
Distribution. Endemic; Fernandina, Isabela.
Bionomics. Phytophagous; host plant: *Scalesia gummifera*; transition and humid forest zone; January, May.

E. helleri Gagné, in Carvalho and Gagné 1968: 174; Froeschner 1985: 35 (*Cyrtopeltis*).
Distribution. Endemic; Santa Fé.
Bionomics. Phytophagous; host plant: *Scalesia helleri*; arid zone; February.

E. modestus (Distant) 1893: 447 (*Neosilia*); Van Duzee 1937: 116 (*Engytatus*); Carvalho and Gagné 1968: 177. Schaefer et al. 1980: 47.
Engytatus geniculata Reuter 1876: 83
Distribution. Adventive; North America, Central America, South America, West Indies; Fernandina, Floreana, Isabela, Pinta, Santiago, Santa Cruz.
Bionomics. Phytophagous on Solanaceae; host plants: tomatoes, potatoes, under *Portulaca*; arid to humid forest and agriculture zones; at lights; February–July.

E. species, probably new; Schaefer et al. 1980: 47
Distribution. Endemic; Pinta.
Bionomics. Phytophagous; evergreen shrub zone?; December.

Genus *Macrolophus* Fiebar 1858

Key to Galápagos species of *Macrolophus*

1a Corium with four fuscous to black spots; 2 in outer apical angle and 2 along inner margin of cuneus; length 2.8–3.1 mm . . . . . *M. punctatus*
1b Corium unicolourous, without above-mentioned spots; length 3.3– 3.5 mm . . . . *M. innotatus*

M. innotatus Carvalho, in Carvalho and Gagné 1968: 167; Froeschner 1985: 36.
Distribution. Endemic; San Cristóbal, Santa Cruz.
Bionomics. Phytophagous?; pampa zone, February–March.

M. punctatus Carvalho, in Carvalho and Gagné 1968: 168; Froeschner 1985: 36.
Distribution. Endemic; Fernandina?, Floreana, Genovesa, Pinzón, Santa Cruz.
Bionomics. Phytophagous?; Arid to pampa and agriculture zones; under bark of *Bursera graveolens*; March–June.

Subfamily Mirinae
Tribe Mirini
Genus *Creontiades* Distant 1883

The seven species of this genus probably descended from four colonizing species.

Key to Galápagos species of *Creontiades*

1a Dorsal surface wholly or in large part reddish brown to black, scutellum mostly or uniformly so 2

1b Dorsal surface pale yellow to yellow-green; scutellum pale, with or without dark spots 4

2a Corium with an irregular subbasal pale area; scutellum wholly fuscous to black; length 5.0–6.0 mm . . . . *C. fuscosus*

2b Corium without a subbasal pale area; scutellum with 2 pale areas subapically. 3

3a Posterior femur with a subapical pale area; pronotum sprinkled with numerous black dots; antennal segment II with strong to weak pale areas subbasally and postmedially; length 5.9–6.1 mm . . . *C. fernandinus*

3b Posterior femur without pale areas; pronotum with few or no black dots; antennal segment II with basal two-thirds pale; length 5.8– 6.3 mm . . . . *C. castaneus*

4a Pale scutellum with at least extreme apex fuscous to black 5

4b Scutellum wholly pale yellow to green. 6

5a Head, pronotum, and scutellum pale yellow with slender red midline, but without dark dots; length 5.2 mm . . . . *C. vittatus*

5b Head, pronotum, and scutellum pale yellow, with numerous fuscous to black dots, but virtually no differentiated midline; length 6.0 mm . . . . *C. punctatus*

6a Posterior tibia with a minute black dot dorsally at base; pronotum pale yellow, usually with a slender fuscous line across base; length 6.2– 6.7 mm. . . . . *C. willowsi*

6b Posterior tibia without basal black dot; pronotum pale green without basal fuscous line; length 5.5 mm. . . . . *C. citrinus*

C. castaneum Van Duzee 1933: 27; Barber 1934: 288; Carvalho and Gagné 1968: 190; Froeschner 1985: 28.
  Distribution. Endemic; Floreana, Isabela, San Cristóbal, Santa Cruz.
  Bionomics. Phytophagous?; humid forest zone; January–June.

C. citrinus Carvalho, in Carvalho and Gagné 1968: 192; Froeschner 1985: 28.
  Distribution. Endemic; Pinta, San Cristóbal, Santa Cruz.
  Bionomics. Phytophagous?; host plant: *Heliotropium curassavicum*?; littoral and arid zone; February–March, June.

C. fernandinus Carvalho, in Carvalho and Gagné 1968: 192; Froeschner 1985: 28.
  Distribution. Endemic; Fernandina, Marchena.
  Bionomics. Phytophagous; arid zone; January–March.

C. fuscosus Barber 1925: 248, 1934: 288; Van Duzee 1937: 115; Carvalho and Gagné 1968: 194; Froeschner 1985: 28.
  Distribution. Endemic; Fernandina, Isabela, Marchena, Pinta, Pinzón, Rábida, Santa Cruz.
  Bionomics. Phytophagous; host plant: *Scalesia gummifera*?; arid to pampa zones; January–July.

C. punctatus Carvalho, in Carvalho and Gagné 1968: 196; Froeschner 1985: 29.
  Distribution. Endemic; Española, Fernandina, Genovesa, Isabela, Marchena, Pinta, Pinzón, Rábida, Santa Cruz.
  Bionomics. Phytophagous?; littoral to evergreen shrub zones; February–June.

C. vittatus Carvalho, in Carvalho and Gagné 1968: 196; Froeschner 1985: 29.
  Distribution. Endemic; Fernandina, Santa Cruz.
  Bionomics. Phytophagous; arid zone; January–February, December.

C. willowsi Van Duzee 1933: 28, 1937: 115; Barber 1934: 288; Carvalho and Gagné 1968: 197; Froeschner 1985: 29.
  Distribution. Endemic; Rábida, Santiago, Santa Cruz.
  Bionomics. Phytophagous; arid zone; January–February.

Genus *Dagbertus* Distant 1904

Twelve species descending from 5 colonizations; see review of the 9 named Galápagos species in Carvalho and Fontes 1983.

Key to Galápagos species of *Dagbertus* (from Froeschner 1985)

1a Corium with 2 red to reddish brown transverse fasciae: basal one extending to midlength of clavus; antennal segment II pale yellow with subbasal and apical parts reddish; length, 3 mm . . . . . . *D. formosus*
1b Corium and antennal segment II not coloured as above 2

2a   Corium white, with numerous brown areas; antennal segment II pale yellow, only extreme apex darkened; length 4.2 mm . . . *D. marmoratus*

2b   Corium variously coloured, when white then without brown areas; antennal segment II either wholly pale or pale only at extreme base . . 3

3a   Pronotum with 4–6 longitudinal brown or red stripes 4

3b   Pronotum without longitudinal stripes 5

4a   Pronotum with 6 longitudinal brown stripes; length, 4.0 mm . *D. darwini*

4b   Pronotum with 4 reddish brown longitudinal stripes; length 3.3 mm
    . . . . *D. figuratus*

5a   Pronotum entirely pale yellow; corium without basal transverse brown band; beak of male not reaching genital capsule 6

5b   Pronotum mostly brown, or at least with a basal transverse brown band . 7

6a   Side of prothorax above anterior leg with or without 2 horizontal brown lines; dorsal surface yellow, variously marked with brown or black on head, pronotum, and cuneus; length 3.5–4.5 mm . . . . . . *D. spoliatus*

6b   Side of prothorax above anterior legs with 2 horizontal red lines; upper surface uniformly yellow except for infuscate membrane; length 3.8–4.1 mm . . . . *D. pallidus*

7a   Pronotum mostly brown with a median, narrow, transverse, yellow area; head almost wholly brownish black; length, 3.4–4.5 mm
    . . . . *D. nigrifrons*

7b   Pronotum mostly pale, with no more than a narrow, subbasal, transverse, brown band; head mostly yellow, sometimes with tylus and median line of interocular area brown 8

8a   Pronotum with subbasal brown band faint; clavus with brown markings restricted to its edges; length, 3.2–4.2 mm . . . . *D. lineatus*

8b   Pronotum with subbasal brown band dark, prominent; clavus with brown markings more extensive; length, 4.5 mm . . . . *D. quadrinotatus*

D. darwini (Butler) 1877: 89 (*Capsus*); Distant 1909: 203; Champion 1924: 260; Barber 1934: 287; Carvalho and Gagné 1968: 206; Froeschner 1985: 30
    Distribution. Endemic; Floreana, Pinzón, Santa Cruz.
    Bionomics. Phytophagous; humid forest zone; host plant: *Scalesia* species?; February–March.

D. figuratus Gagné, in Carvalho and Gagné 1968: 217; Froeschner 1985: 30.
    Distribution. Endemic; Santa Cruz.
    Bionomics. Phytophagous?; arid zone; February–March.

D. formosus Carvalho, in Carvalho and Gagné 1968: 207; Froeschner 1985: 31.
    Distribution. Endemic; Santa Cruz.
    Bionomics. Phytophagous?; arid zone; January–February.

D. lineatus Gagné, in Carvalho and Gagné 1968: 216; Froeschner 1985: 31.

Distribution. Endemic; Fernandina, (not Floreana).
Bionomics. Phytophagous?; pampa zone; February.

D. marmoratus Carvalho, in Carvalho and Gagné 1968: 209; Froeschner 1985: 31.
Distribution. Endemic; Floreana.
Bionomics. Phytophagous?; humid forest zone?; February.

D. nigrifrons Gagné, in Carvalho and Gagné 1968: 214; Froeschner 1985: 31.
Distribution. Endemic; Floreana.
Bionomics. Phytophagous?; humid forest zone; February.

D. pallidus Gagné, in Carvalho and Gagné 1968: 212; Froeschner 1985: 31
Distribution. Endemic; Santa Cruz.
Bionomics. Phytophagous?; arid zone; February.

D. quadrinotatus (Walker) 1873: 113 (*Capsus*); Butler 1877: 89 (*Capsus*); Champion 1924: 260; Barber 1934: 288; Carvalho and Gagné 1968: 210; Froeschner 1985: 31.
Distribution. Endemic; Floreana, San Cristóbal, Santiago, Santa Cruz.
Bionomics. Phytophagous?; arid to humid forest and agriculture zones; January–May.

D. spoliatus (Walker) 1873: 112 (*Capsus*); Butler 1877: 89 (*Capsus*); Champion 1924: 260; Barber 1934: 288; Carvalho and Gagné 1968: 212; Froeschner 1985: 32.
Distribution. Endemic; Floreana, Santiago.
Bionomics. Phytophagous?; humid forest; February.

D. species A. Carvalho, in Carvalho and Gagné 1968: 218.
Distribution. Endemic; Pinzón.
Bionomics. Phytophagous?; humid forest zone; February.

D. species B. Carvalho, in Carvalho and Gagné 1968: 218.
Distribution. Endemic; San Cristóbal.
Bionomics. Phytophagous?; agriculture zone; February.

D. species C. Carvalho, in Carvalho and Gagné 1969: 218.
Distribution. Endemic; San Cristóbal
Bionomics. Phytophagous?; February.

Genus *Galapagomiris* Carvalho 1968 (endemic genus)

G. longirostris Carvalho, in Carvalho and Gagné 1968: 188; Froeschner 1985: 32.
Distribution. Endemic; Rábida, Santa Cruz.
Bionomics. Phytophagous?; arid zone; January, May.

Genus *Horcias* Distant 1884

H. chiriquinus Carvalho and Gagné 1968: 198 (not Distant 1884: 278).
   Distribution. Adventive?; Mexico to Bolivia; Santa Cruz.
   Bionomics. Phytophagous; agricultural zone; January–February, April,
   July.

Genus *Polymerus* Hahn 1831

P. vegatus (Van Duzee) 1933: 28 (*Poeciloscytus*); Butler 1877: 89; Champion
1924: 260 (*Capsus*); Barber 1934: 288 (*Polymerus*); Carvalho and Gagné 1968:
200; Froeschner 1985: 33 (as *P.*, and *C. nigritulus* Walker).
   Distribution. Endemic; Española, Fernandina, Floreana, Isabela, Marchena,
   Pinta, Santa Cruz, Santiago.
   Bionomics. Phytophagous?; Arid to pampa zones; host plants: *Jaegeria
   hirta*, *Portulaca* species; January–June.

Genus *Taylorilygus* Leston 1952

T. apicalis (Fieber) 1861: 275 (*Lygus*); Carvalho and Gagné 1968: 203;
Froeschner 1985: 33 (as *Taylorilygas* or *Phytocoris pallidulus* Blanchard).
   Distribution. Adventive?; cosmopolitan; Floreana, San Cristóbal, Santa
   Cruz.
   Bionomics. Phytophagous?; agriculture zone; February–April, September.

Tribe Stenodemini
Genus *Dolichomiris* Reuter 1882

D. linearis Reuter 1882: 29; Carvalho and Gagné 1968: 183; Schaefer et al.
1980: 48; Froeschner 1985: 33.
   Distribution. Indigenous?; cosmopolitan; (present in 1830's in Darwin's col-
   lections); Fernandina, Isabela, Pinta, Santa Cruz, Santiago.
   Bionomics. Phytophagous?; Arid and inversion zones; swept from grass;
   March–June.

Genus *Trigonotylus* Fieber 1858

T. lineatus (Butler) 1873: 89 (*Miris*); Carvalho and Gagné 1968: 185.
   Distribution. Endemic?; Floreana, Isabela, Pinta, San Cristóbal, Santiago.
   Bionomics. Phytophagous; arid and transition forest; March.

Superfamily Miroidea
Family Tingidae
The Lace Bugs

These occur in fine vegetation such as moss and leaves and the Galápagos species seem to feed on leaf tissue, usually on the underside of leaves; three colonizations may account for the six species.

Key to Galápagos genera of Tingidae

1a Pronotum anteriorly with a large, inflated hood projecting over and beyond head; posterior margin of pronotum produced rearward as a long, acutely angled process. . . . . *Corythaica*

1b Pronotum anteriorly without a hood, head wholly visible in dorsal view; posterior margin of pronotum transverse 2

2a Head top and sides ornamented with about 9 horn-like projections
. . . . *Phatnoma*

2b Head not ornamented with horns, with small low ridges only *Teleonemia*

Subfamily Cantacaderinae
Genus *Phatnoma* Fieber 1844

Key to Galápagos species of *Phatnoma*

1a Costal area uniformly expanded and, except at basal fourth, with two rows of cells; length 2.6 mm . . . . *P. biordinatum*

1b Costal area uniformly expanded with three rows of cells for virtually full length; length 3.9 mm . . . . *P. eremaea*

P. biordinatum Froeschner 1976: 183; Froeschner 1985: 55.
  Distribution. Endemic; Santa Cruz.
  Bionomics. Phytophagous; transition zone, under bark; January–February. Both brachypterous and macropterous individuals are known. Previously known only from holotype.

P. eremaeum Drake and Froeschner 1967: 83; Froeschner 1976: 184, 1985: 55.
  Distribution. Endemic; Santa Cruz.
  Bionomics. Phytophagous; transition, humid forest and agricultural zones. January–July.

Subfamily Tinginae
Tribe Tingini
Genus *Corythaica* Stål 1873

Key to Galápagos species of *Corythaica*

1a Lateral expansion of pronotum, in dorsal view, concave in front of submedian angular expansion; dorsal surface variously marked with fuscous areas. 2

1b Lateral expansion of pronotum, viewed from above, not abruptly concave in anterior half; dorsum milky white, with no fuscous markings except sometimes on veins; length 2.4–2.6 mm. . . . . . *C. darwiniana*

2a Anterior half of pronotal hood tumidly swollen similar to basal half, scarcely narrowing to the broadly blunt apex; lateral carinae of pronotal disc not more than one-third as high as median carina; length 2.3–2.6 mm. . . . . *C. wolfiana*

2b Anterior half of pronotal hood compressed, gradually tapering to very acute apex; lateral carinae on pronotal disc two-thirds to fully as high as median carina; length 2.5–2.7 mm . . . . *C. cytharina*

C. cytharina (Butler) 1877: 9 (*Monanthia*); Barber 1934: 286; Drake and Froeschner 1967: 84. Froeschner 1976: 182, 1985: 55.
> *Monanthia renormata* Barber 1925: 251.
> Distribution. Endemic; Baltra, Daphne Major, Española, Fernandina, Floreana, Genovesa, Isabela, Marchena, Mosquera, Pinta, Pinzón, Rábida, San Cristóbal, Santa Cruz, Santiago.
> Bionomics. Phytophagous; host plants: (On 5 plant families) *Cryptocarpus pyriformis*, *Lycopersicon esculentum* (pest); *Portulaca* species, *Scalesia affinis*, *S. gummifera*, *Side rhombifolia*; littoral to evergreen shrub zone; January–July.

C. darwiniana Drake and Froeschner 1967: 89; Froeschner 1976: 182, 1985: 56.
> Distribution. Endemic; Darwin.
> Bionomics. Phytophagous; arid zone; host plant: *Alternanthera*?; January.

C. wolfiana Drake and Froeschner 1967: 87; Froeschner 1976: 182, 1985: 56
> Distribution. Endemic; Isabela, Pinzón, Wolf.
> Bionomics. Phytophagous; host plant: *Scalesia* species; arid to pampa zones; February, June.

Genus *Teleonemia* Costa 1864

T. prolixa Stål 1858: 65; New archipelago record.
Distribution. Indigenous; widespread New World tropics; Floreana, San Cristóbal, Santa Cruz.
Bionomics. Phytophagous; arid to pampa and agriculture zones; FIT; March–May.

Superfamily Naboidea
Family Nabidae
The Nabid Bugs or Damsel Bugs
Subfamily Nabinae
Genus *Nabis* Latreille 1802

Key to Galápagos species of *Nabis*

1a Brachypterous, wings not or only slightly longer than pronotum. . . . 2
1b Macropterous, wings more than twice as long as pronotum; length, 7.5–8 mm. . . . . *N. consimilis*
2a Tibia with light brown spots, base and apex brown; length, 4.9–6.5 mm; ocelli normal . . . . *N. galapagoensis*
2b Tibia with distinct brown spots, base and apex not brown; length, 4.6–5.3 mm; ocelli strongly reduced, only flattened reddish traces present
 . . . . *N. reductus*

N. consimilis (Reuter) 1912: 23 (*Reduviolus*); Kerzhner 1968: 85; Schaefer et al. 1980: 47; Froeschner 1985: 39
 N. punctipennis Heidemann 1901: 366 (not Blanchard); Barber 1925: 251; Kerzhner 1968: 85.
 Distribution. Indigenous; Ecuador, Peru; Baltra, Española, Fernandina, Floreana, Isabela, Pinta, Pinzón, San Cristóbal, Santa Cruz, Santiago.
 Bionomics. Predator on insects; littoral to pampa zone and urban and agricultural habitats; at lights; February–September.

N. galapagoensis Kerzhner 1968: 86; Froeschner 1985: 40
 Distribution. Endemic; Floreana, Isabela, Santa Cruz, Santiago.
 Bionomics. Predator on insects; humid forest to pampa zone, January–December.

N. reductus Kerzhner 1968: 90; Froeschner 1985: 40
 Distribution. Endemic; San Cristóbal, Santiago.
 Bionomics. Predator on insects; humid forest to pampa zone; from weeds, under stones, from lichens on *Psidium guajava*; February–April, June.

Superfamily Cimicoidea
Family Anthocoridae
The Minute Pirate Bugs

Key to Galápagos genera of Anthocoridae

1a Greatest diameters of antennal segments II, III and IV equal; setae on antennal segments III and IV less than twice as long as diameter of supporting segment. . . . . *Orius*

1b Antennal segments III and IV distinctly thinner than II and with long erect setae, most of which are at least twice as long as diameter of supporting segment. 2

2a In dorsal view, lateral margin of pronotum laminately expanded for full length, width of expansion more than half as wide as first antennal segment. . . . . *Nidicola*

2b In dorsal view, lateral margin of pronotum not expanded 3

3a Clavus and scutellum with numerous distinct punctures which are sometimes darkened 4

3b Clavus and scutellum without punctures (although short setae may suggest punctures) 5

4a Lateral margin of pronotum and costal margin with crowded long setae projecting laterally and horizontally; pronotal calli laterally reaching lateral pronotal margins, not limited by a longitudinal submarginal impression . . . . *Lasiochilus*

4b Lateral margin of pronotum and costal margin without long horizontal setae; pronotal calli abruptly delimited laterally . . . . . . *Amphiareus*

5a Pronotal surface smoothly convex, without a prominent transverse impression posterior to calli; corium whitish hyaline, colour contrasting sharply with polished reddish brown of head, pronotum and scutellum . . . . *Xylocoris*

5b Pronotal disc with a distinct transverse impression posterior to calli; corium not whitish hyaline 6

6a Metapleural scent trough (peritreme) short, curved caudad, attaining middle of posterior margin of dull roughened evaporating surface . *Alofa*

6b Metapleural scent trough long, gently curved anteriorly, more or less paralleling posterior margin of dull roughened evaporating area and reaching its lateral margin . . . . *Cardiasthethus*

Subfamily Anthocorinae
Tribe Oriini
Genus *Orius* Wolff 1811

O. tristicolour (White) 1879: 145 (*Triphleps*); Herring 1966: 130; Froeschner 1985: 8

Distribution. Indigenous; North America, Central America, South America, West Indies; Española, Fernandina, Isabela, Marchena, Pinta, Pinzón, Santa Cruz, Santiago.

Bionomics. Insect predator or pollen feeder?; littoral to pampa zones; March–August.

Subfamily Lasiochilinae
Genus *Lasiochilus* Reuter 1871

L. pallidulus Reuter 1871: 562; Herring 1966: 128; Schaefer et al. 1980: 46; Froeschner 1985: 8

Distribution. Indigenous; North America, Central America, West Indies; Isabela, Marchena, Pinta, Santa Cruz, Santiago.

Bionomics. Insect predator; littoral to pampa zones; at lights, under *Sesuvium*; January–July.

Subfamily Lyctocorinae
Tribe Cardiastethini
Genus *Alofa* Herring 1976

A. sodalis (White) 1878: 372 (*Cardiastethus*); Herring 1966: 127 (*Buchananiella*); Froeschner 1985: 9

Distribution. Adventive?; almost cosmopolitan; Isabela, Santa Cruz.

Bionomics. Insect predator; transition to evergreen shrub zone; January, April.

Genus *Amphiareus* Distant 1904

A. constrictus (Stål) 1860: 44 (*Xylocoris*); Herring 1966: 127 (*Buchananiella*); Froeschner 1985: 9.

Distribution. Adventive?; almost cosmopolitan; Isabela, Santa Cruz.

Bionomics. Insect predator; littoral to pampa zones; March, May–July.

Genus *Cardiastethus* Fieber 1860

C. limbatellus (Stål) 1860: 44 (*Xylocoris*); Herring 1966: 127; Froeschner 1985: 9.

Distribution. Indigenous?; Central America, South America; Baltra, Floreana, Isabela, Santa Cruz.

Bionomics. Predator on insects; littoral and humid forest zones; January–
June.

Tribe Scolopini
Genus *Nidicola* Harris and Drake 1941

N. mazda Herring 1966: 129; Froeschner 1985: 10.
  Distribution. Endemic; Fernandina, Isabela, Marchena, Rábida, Santa Cruz,
    Wolf.
  Bionomics. Predator on insects; littoral to transition zones; January–
    June.

Tribe Xylocorini
Genus *Xylocoris* Dufour 1831

Key to Galápagos species of *Xylocoris*

1a Body larger, about 3 mm long; pronotum and scutellum darker brown;
    wings more white . . . . *X. sordidus*
1b Body smaller, about 2 mm long; pronotum and scutellum lighter brown;
    wings more yellowish . . . . *X.* undetermined

X. sordidus (Reuter) 1871: 560 (*Piezostethus*); Herring 1966: 130; Froeschner
1985: 10
  Distribution. Indigenous; North America, Central America, South America,
    West Indies; Baltra, Isabela, Pinta, San Cristóbal, Santa Cruz, Santa Fé.
  Bionomics. Predator on insects; littoral to humid forest, and urban agricul-
    ture zones; January–June.

Xylocoris undetermined
  Distribution. Adventive?; San Cristóbal.
  Bionomics. Predator on insects; arid zone and village; March; malaise and
    light traps.

Family Cimicidae
The Bed Bugs
Genus *Cimex* Linnaeus 1758

C. lectularius Linnaeus 1758: 441; Barber 1934: 281; Froeschner 1985: 11; the
human bed bug.
  Distribution. Adventive; cosmopolitan; Floreana.
  Bionomics. Ectoparasitic blood feeder on humans; the record is based only
    on a single 1925 collection; December.

Suborder Pentatomorpha
Superfamily Pentatomoidea
Family Cydnidae
The Burrowing Bugs

Froeschner (1960: 438, 1968: 192, 1985: 15) has listed *Melanaethus subglaber* (Walker) of western USA and Mexico from the Galápagos, but indicated that it may be a mislabeled record. Because there are no other valid records, I here delete it from the Galápagos list, but retain it in the following key in case it does actually occur in Galápagos. The thousands of cydnids that have now been collected and identified as *Dallasiellus murinus* strengthen the suspicion that the *M. subglaber* record is an error.

Key to Galápagos genera of Cydnidae

1a In ventral view ostiolar pore visible, peritreme laterad of it with a polished lobe. . . . . *Melanaethus*
1b In ventral view ostiolar pore concealed, peritreme in no part polished
. . . . *Dallasiellus*

Subfamily Cydninae
Genus *Dallasiellus* Berg 1901

D. murinus (Van Duzee) 1933: 26 (*Geotomus*); Barber 1934: 282 (*Geotomus*); Froeschner 1960: 616, 1968: 192, 1985: 14; Schaefer et al. 1980: 44.
  Distribution. Indigenous; Ecuador mainland; Campeón, Fernandina, Floreana, Isabela, Pinta, Pinzón, Rábida, San Cristóbal, Santa Cruz, Santiago.
  Bionomics. Phytophagous on plant roots; littoral to pampa and agriculture zones; at lights, in litter, in cave; January–December.

Family Pentatomidae
The Stink Bugs

Keys to Galápagos genera of Pentatomidae

1a Body elongate and slender, nearly linear, about 4 times as long as broad
. . . . *Mecidea*
1b Body relatively broad and ovoid, not over 3 times as long as broad . . 2
2a Humeri of prothorax projecting laterally as an elongate blunt or sharp process 3
2b Humeri not projecting laterally as a spine or long lobe 4

3a Humerus projecting laterally as a prominent, apically unevenly branched, stout spine; juga not surpassing apex of clypeus . . . . . *Alcaeorrhynchus*

3b Humerus projecting as a sharp, tapering spine; juga distinctly surpassing apex of clypeus as a long, narrow, finger-like projection . . . . *Loxa*

4a Lateral margin of pronotum, at least on apical half, with distinct, coarse serrations; buccalae less than half as long as rostral segment I . . . . *Podisus*

4b Lateral margin of pronotum entire, not serrate 5

5a Venter of abdomen mediobasally projecting forward toward posterior coxae . . . . *Acrosternum*

5b Venter of abdomen without a mediobasal projection 6

6a Lateral margin of head broadly (in part as wide as or wider than an eye) laminately expanded, its edge with a prominent angled projection a short distance anterior to eye. . . . . *Trincavellius*

6b Lateral margin of head neither laminately expanded nor with preocular projection . . . . *Thyanta*

Subfamily Asopinae
Genus *Alcaeorrhyncus* Bergroth 1891

A. grandis (Dallas) 1851: 91 (*Canthecona*); Barber 1934: 283; Froeschner 1985: 41.

> Distribution. Adventive; North America, Central America, South America; Española, Floreana, Isabela, Santa Cruz.
> Bionomics. Phytophagous? or predator?; arid, transition, pampa and inversion zones. March–June.

A. species 2.

> Distribution. Indigenous?; Isabela.
> Bionomics. Predator; pampa zone; May.

Genus *Podisus* Herrich-Schaeffer 1851

Key to Galápagos species of *Podisus*

1a. Pronotal humeral angles rounded, short and blunt, without conspicuous secondary posterior tooth; abdominal sternites with sparse dark areas and few dark punctures . . . . *P. sordidus*

1b. Pronotal humeral angles long and pointed, with conspicuous smaller tooth behind major angle; abdominal sternites with large dark areas and many dark punctures . . . . *P.* sp. 2.

P. sordidus (Stål) 1859: 221 (*Arma*); Barber 1925: 241; Heidemann 1901: 364; Barber 1934: 283; Schaefer et al. 1980: 45.

> Distribution. Indigenous?; Peru; Española, Fernandina, Floreana, Isabela, Marchena, Pinta, Pinzón, San Cristóbal, Santa Cruz, Santiago, Seymour.
> Bionomics. Predator on insects?; arid to pampa zone and urban areas; in litter, around boulders, at light; March–June, December.

P. species 2?

> Distribution. Indigenous?; Isabela, Santa Cruz, Santiago.
> Bionomics. Predator on insects?; transition and pampa zones; April–May, December.

Subfamily Discocephalinae
Genus *Trincavellius* Distant 1900 (endemic genus)

T. galapagoensis (Butler) 1877: 88 (*Sciocoris*); Champion 1924: 260; Barber 1934: 282; Froeschner 1985: 42.

> Distribution. Endemic; Española, Floreana.
> Bionomics. Phytophagous?; arid zone; May, December.

Subfamily Pentatominae
Tribe Mediceini
Genus *Mecidea* Dallas 1851

M. species. New genus record for the archipelago.

> Distribution. Indigenous?; Rábida.
> Bionomics. Phytophagous; littoral-arid zone; March; known from two specimens; sweeping in grass; an unusually narrow species that probably feeds on grasses. The tribe and genus are not reported from mainland Ecuador (Froeschner 1981).

Tribe Pentatomini
Genus *Acrosternum* Fieber 1860

Key to Galápagos species of *Acrosternum*

1a Dorsal punctures widely separated, usually by a space at least as great as width of antennal segment II; abdominal spiracles not surrounded by a callus; length 10–12.5 mm . . . . *A. viridans*

1b Dorsal punctures close-set, spaces between them much less than width of antennal segment II; abdominal spiracles placed on a yellowish callus; length 10.6–13.7 mm . . . . *A. ubicum*

A. ubicum Rolston 1983: 135; Froeschner 1985: 42
    Distribution. Indigenous?; West Indies, South America; Española, Floreana, Isabela, San Cristóbal, Santa Cruz.
    Bionomics. Phytophagous?; arid zone to pampa and urban; March–June.

A. viridans (Stål) 1859: 228 (*Rhaphigaster*); Uhler 1889: 194; Heidemann 1901: 365; Van Duzee 1937: 113 (all *Nezara*); Barber 1925: 241, 1934: 282; Rolston 1983: 128; Schaefer et al. 1980: 44; Froeschner 1985: 42.
    Distribution. Indigenous?; Panama to Peru; Fernandina, Floreana, Isabela, Pinta, Pinzón, San Cristóbal, Santa Cruz, Santiago.
    Bionomics. Phytophagous; arid to pampa zones; at lights; February–June, August.

Genus *Loxa* Amyot and Serville 1843

L. viridis (Palisot de Beauvois) 1805: 111 (*Pentatoma*); Froeschner 1985: 43
P. picticornis Horvath 1925: 312; Eger 1978: 243.
    Distribution. Adventive?; Florida and Texas to Brazil and Argentina; Floreana, Isabela, Santa Cruz.
    Bionomics. Phytophagous?; arid to humid forest zone; March–June.

Genus *Thyanta* Stål 1860

Key to Galápagos species of *Thyanta*

1a   Length, 9.6–10 mm; pronotum with humerus forming a short, slightly acute angle projecting anterolaterally, the margin posterior to angle continuing line of costal margin; each pronotal callus with a black spot at inner end; post-spiracular black spots of abdomen distinct, placed mesad of a line connecting spiracle . . . . *T. setigera*
1b   Length, 7–7.1 mm; pronotum with humerus forming at obtuse angle directed (but not projecting) laterad, margin posterior to angle oblique, not continuing line of costal margin; pronotal callus not marked with black . . . . *T. similis*

T. setigera Ruckes 1957: 179; Froeschner 1985: 43.
    T. perditor Heidemann 1901: 365 (not Fabricius, 1794); Barber 1934: 282; Van Duzee 1937: 112.
    Distribution. Endemic; Fernandina, Floreana, Gardner at Floreana, Isabela, Jervis, Rábida, San Cristóbal, Santa Cruz, Santiago, Seymour.
    Bionomics. Phytophagous?; arid zone; at UV light; March–June.

T. similis Van Duzee 1933: 26; Barber 1934: 282; Froeschner 1985: 44.
    Distribution. Endemic; Floreana, Genovesa, Isabela, Pinta, Rábida, Santa Cruz.
    Bionomics. Phytophagous; littoral to transition forest; March–June.

Superfamily Lygaeoidea
Family Berytidae
The Stilt-legged Bugs
Genus *Metacanthus* Costa 1844

M. tenellus Stål 1859: 236; Henry 1997*a*: 129.

M. galapagensis (Barber) 1934: 284 (*Aknisus*); Froeschner 1985: 10.

Jalysus (*Metacanthus*) *tenellus* (Heidemann 1901: 366).

Distribution. Indigenous; Florida and Texas, West Indies, Central America to Argentina and Chile; Floreana, Santa Cruz, San Cristóbal.

Bionomics. Probably phytophagous on plant sap, or predator on soft-bodied insects; arid, humid forest, and agriculture zones; *on Datura*; March–July.

Family Lygaeidae *sensu lat.*
The Seed Bugs

There is substantial diversity in the feeding habitats of this large family. Most are predators on mature seeds, but some may be sap suckers or predators. The higher classification followed here is that of Slater and O'Donnell (1995), but it should be noted that Henry (1997*b*) has presented evidence to split the family into several families.

Key to Galápagos genera of Lygaeidae

1a Head behind the eyes prolonged and gradually narrowing posteriorly, forming a distinct neck; eyes separated from anterior margin of pronotum by a space greater than the horizontal diameter of an eye
. . . . *Heraeus*

1b Head not prolonged into a neck; eyes touching anterior margin of pronotum or separated therefrom by a space distinctly less than the horizontal diameter of an eye 2

2a Lateral margin of pronotum expanded as a narrow (but distinct) ledge from anterolateral angle to posterolateral angle . . . . *Tempyra*

2b Lateral margin of pronotum vertically convexly rounded, without a ledge 3

3a Pronotum anteriorly with a narrow, marginal collar set off posteriorly by a sharply incised, transverse, subapical line extending to and around the lateral margins 4

3b Pronotum with neither an anterior collar nor a sharp, transverse, subapical groove 6

4a Corium with long, erect, pale setae, and sharply delimited, white to yellow, oval spot in inner apical angle . . . . . . *Pseudopachybrachius*

4b Corium without erect setae and without an oval pale spot in inner apical angle. 5

5a Clavus with three parallel rows of punctures, under 4 mm in length; anterior lobe of pronotum and scutellum without long upstanding setae . . . . *Prytanes*

5b Clavus with more than three parallel rows of punctures; over 5 mm in length; anterior lobe of pronotum and scutellum with long upstanding setae . . . . *Neopamera*

6a Clavus opaque on basal third, hyaline on apical two-thirds; mesocorium nearly entirely hyaline, along middle vein with a row of coarse, darkened punctures each giving rise to long seta. . . *Cymoninus*

6b Clavus uniformly opaque for full length; mesocorium without a row of setigerous punctures along middle vein 7

7a Antenniferous tubercle with lateroapical angle narrowly and acutely (finger-like) produced anteriorly; costal margins abruptly diverging at above level of midlength of scutellum. . . . . *Darwinysius*

7b Antenniferous tubercle not produced; costal margins straight or weakly diverging at above level of basal third of scutellum. 8

8a Hemelytron narrow, leaving connexivum exposed for nearly full length; costal margin paralleling stem of branched corial vein. . . *Neortholomus*

8b Hemelytron broader, covering connexivum for full length; costal margin gradually diverging from stem of branched corial vein 9

9a Costal margin of corium and lateral margin of abdomen without a stridulitrum . . . . *Nysius*

9b Costal mrgin of corium and lateral margin of abdomen with a stridulitrum . . . . *Xyonysius*

Subfamily Orsillinae
Tribe Orsillini
Genus *Neortholomus* Hamilton 1983

N. usingeri (Ashlock) 1972: 91 (*Ortholomus*); Hamilton 1983: 226; Froeschner 1985: 21

   Distribution. Endemic; Fernandina, Floreana, Isabela, Pinzón, Rábida, San Cristóbal, Santa Cruz.

   Bionomics. Phytophagus; host plants: *Cordia* species, *Hypericum pratense*, *Verbena* species; arid to pampa and inversion zones; February, April–July.

Notes. *N. usingeri* is the sister species of *N. gibbifer* (Berg) of northern Argentina, southern Brazil and north central Chile and these are the sister group of *N. procerodorus* Hamilton of Andean Peru.

Tribe Metrargini
Genus *Darwinysius* Ashlock 1967 (endemic genus)

Phylogenetic analysis (Asquith 1994) suggests that this genus is a sister to *Robinsonocoris* which is endemic in Juan Fernandez Islands, and that these form a clade with *Balionysius* and *Coleonysius* of South America. This lineage also colonized Hawaii and underwent extensive differentiation there. The two genera in this tribe in the Galápagos represent two separate colonizations (Asquith 1994).

Key to Galápagos species of *Darwinysius*

1a Body exceptionally setose; rare . . . . *D.* n.species
1b Body normally setose; usually common 2
2a Males more than 3.2 mm long, females more than 3.8 mm long; dorsal vestiture including only short, appressed setae; interocular surface of head (in lateral view) convexly raised above top of eyes *D. wenmanensis*
2b Males less than 3.1 mm long, females less than 3.5 mm; dorsal vestiture a mixture of suberect and appressed setae; interocular surface of head not raised above top of eyes . . . . *D. marginalis*

D. marginalis (Dallas) 1852: 556 (*Nysius*); Dohrn 1859: 33; Stål 1874: 122; Butler 1877: 88; Heidemann 1901: 366; Champion 1924: 260; Barber 1925: 245, 1934: 285; Van Duzee 1937: 114; Usinger and Ashlock 1966: 230 (all as *Nysius*).
  Cymus galapagensis Stål 1859: 252; Ashlock 1972: 95; Schaefer et al. 1980: 46; Froeschner 1981: 42, 1985: 20.
  Distribution. Endemic; Bartolomé, Beagle, Daphne Major, Fernandina, Floreana, Isabela, Mosquera, Pinta, Pinzón, Rábida, Santa Cruz, San Cristóbal, Santiago, Sombrero Chino.
  Bionomics. Seed predator; host plants: *Portulaca* species, *Euphorbia viminea*; arid to pampa zone and agricultural zone; March–September.

D. wenmanensis Ashlock 1972: 95; Usinger and Ashlock 1966: 233 (*Nysius* species); Froeschner 1985: 20.
  Distribution. Endemic; Wolf.
  Bionomics. Seed predator?; under *Portulaca*; arid zone; January, September.

D. new species
  Distribution. Endemic; Española, Rábida.

Bionomics. Seed predator; littoral and arid zones. An exceptionally setose species, distinct from the other known species; known from only 3 specimens; April and June.

Genus *Xyonysius* Ashlock and Lattin 1963

X. naso (Van Duzee) 1933: 27 (*Nysius* (*Ortholomus*)); Barber 1934: 285 (*Ortholomus*); Usinger 1941: 31 (*Nysius*); Usinger and Ashlock 1966: 235; Ashlock 1972: 97; Froeschner 1985: 21.
   Distribution. Endemic; Floreana, Fernandina, Isabela, Pinzón, Santa Cruz, Santa Fé.
   Bionomics. Phytophagous seed predator?; arid to evergreen shrub zones; host plants: *Scalesia affinis*, *S. gummifera*, *S. helleri*; May–July. The closest relative to this species seems to be *X. acticola* Baranowski and Slater (1997) of the Turks and Caicos Islands, at the southern end of the Bahamas Island group.

Tribe Nysiini
Genus *Nysius* Dallas 1852

N. usitatus Ashlock 1972: 89; Froeschner 1985: 21.
   Distribution. Endemic; Fernandina, Floreana, Isabela, Santa Cruz.
   Bionomics. Phytophagous?; littoral to pampa zones; found under *Portulaca*; March–June.

Subfamily Cyminae
Tribe Ninini
Genus *Cymoninus* Breddin 1907

C. notabilis (Distant) 1882: 191 (*Ninus*); Ashlock 1972: 97; Froeschner 1985: 19.
   Distribution. Indigenous?; extreme southern USA, Central America, South America, West Indies; Floreana, Isabela (Sierra Negra), San Cristóbal, Santa Cruz, Santiago.
   Bionomics. Phytophagous on plant juices; arid zone to pampa; host plant: *Cyperus confertus*; March–July.

Subfamily Rhyparochrominae
Tribe Udeocorini
Genus *Tempyra* Stål 1874

T. biguttula Stål 1874: 157; Froeschner 1985: 23.
   Distribution. Indigenous?; USA, Mexico; San Cristóbal, Santa Cruz.
   Bionomics. Phytophagous; arid zone; March, May.

Tribe Myodochini
Genus *Heraeus* Stål 1862

H. pacificus Barber 1925: 245, 1934: 286; Slater 1964: 1083; Ashlock 1972: 101; Parkin et al. 1972: 102; Schaefer et al. 1980: 46; Froeschner 1985: 22.
> Distribution. Endemic; Fernandina, Floreana, Isabela, San Cristóbal, Santa Cruz, Santiago.
> Bionomics. Phytophagous?; host plant? *Jaegaria hirta*; arid to pampa zones; at light; January–December.

Genus *Neopamera* Harrington 1980

N. insularis (Barber) 1925: 246, 1934: 266 (*Orthaea*); Ashlock 1972: 98; Schaefer et al. 1980: 46 (all *Pachybrachius*); Froeschner 1985: 22.
> Distribution. Endemic; Baltra, Fernandina, Floreana, Isabela, Marchena, Pinta, Pinzón, Plazas, Rábida, Santa Cruz, Santiago.
> Bionomics. Phytophagous?; in humus of littoral to pampa and agriculture and urban zones; at lights; February–December.

Genus *Prytanes* Distant 1893

P. confusus (Barber) 1953: 21 (*Exptochiomera*); New archipelago record.
> Distribution. Adventive; Texas, Florida, Cuba, Mexico, Central America, Ecuador (Guayaquil); Santa Cruz.
> Bionomics. Phytophagus; arid zone; March and April; uv light at CDRS; first records in 1992.

P. undescribed species.
> Distribution. Endemic; San Cristobal.
> Bionomics. Phytophagous; humid forest on summit of Cerro San Joachin at 730 m, in pitfall trap; with reduced flight wings; generic placement not certain; February

Genus *Pseudopachybrachius* Malipatil 1978

P. nesovinctus (Ashlock) 1972: 98 (*Pachybrachius*); Schafer et al. 1980: 46; Zheng and Slater 1984.
> Distribution. Endemic; Fernandina, Isabela, San Cristóbal, Santa Cruz, Santiago.
> Bionomics. Phytophagous; arid to pampa and agriculture zones; February–December.

Superfamily Pyrrhocoroidea
Family Pyrrhocoridae
The Cotton Stainers
Genus *Dysdercus* Guerin-Meneville 1831

Key to Galápagos species of *Dysdercus*

1a Abdominal segments ventrally red with posterior margins white to yellow; length, 10–16 mm . . . . *D. concinnus*

1b Abdominal segments ventrally either wholly red or red with transverse fuscous bands; length, 10.5–17 mm . . . . *D. lunulatus*

D. concinnus Stål 1861: 198; Barber 1925: 248, 1934: 286; Froeschner 1985: 45
   Distribution. Adventive?; southern USA, Central America, north western South America; Santa Cruz (based on single Barber 1925 record).
   Bionomics. Phytophagous on fruits or seeds?; arid zone?; April?; I have no new records.

D. lunulatus Uhler 1861: 24; Doesburg 1968: 88; Froeschner 1985: 45.
   Distribution. Adventive?; southern Mexico to north western South America; San Cristóbal.
   Bionomics. Phytophagous or fruits or seeds?; arid zone; April?; I have no new records.

Superfamily Coreoidea
Family Coreidae
The Leaf-footed Bugs
Subfamily Coreinae
Tribe Syromastini
Genus *Anasa* Amyot and Serville 1843

Key to Galápagos species of *Anasa*

1a Size larger, about 10 mm long; first antennal segment long and thin, of uniform diameter; projection of antennal tubercle longer and thinner
. . . . *A. sorbutica*

1b Size smaller, about 8 mm long; first antennal segment shorter, swollen except in basal 1/4; projection of antennal tubercle shorter and stouter
. . . . *A. mimetica*

A. sorbutica (Fabricius) 1775: 706 (*Cimex*); Dohrn 1859: 30; Stål 1870: 197; Butler 1877: 88; Heidemann 1901: 365; Champion 1924: 260; Barber 1934: 283; Froeschner 1985: 12 (all as *A. obscura* Dallas); Brailovsky 1985: 179.
> Distribution. Adventive; widespread from southern USA, through the West Indies and Central America to Brazil and Argentina. Isabela, Pinta, San Cristóbal, Santa Cruz.
> Bionomics. Phytophagous; arid to pampa and urban and agriculture zones; March–July, and November; an agricultural pest on many crops, especially cucurbits.

A. mimetica Brailovsky 1985: 154. New archipelago record.
> Distribution. Adventive; Ecuador; Floreana.
> Bionomics. Phytophagous; arid and agriculture zones, April; previously known from only Ambato, Ecuador (Brailovsky 1984: 154).

Family Rhopalidae
The Scentless Plant Bugs

Key to Galápagos genera of Rhopalidae

1a Posterior femora distinctly thicker than middle or anterior femora and with a row of thick, prominent spines ventrally; lateral margin of pronotum with a row of prominent, projecting angulations
. . . . *Harmostes*

1b Posterior femora not thicker than middle and anterior femora, unarmed; lateral margin of pronotum without a row of projections 2

2a Corium opaque, black, with abundant white decumbent setae interspersed with coarse black bristles that make the surface appear speckled . . . . *Jadera*

2b Corium hyaline, bare 3

3a Pronotum anterior to transverse impressed stripe polished and not punctate . . . . *Liorhyssus*

3b Pronotum anterior to transverse impressed stripe not smooth and polished, always with numerous coarse punctures present . . . . . *Arhyssus*

Subfamily Rhopalinae
Tribe Harmostini
Genus *Harmostes* Burmeister 1835

H. disjunctus Barber 1925: 241; 1934: 283; Göllner-Scheiding 1978: 272; Schaefer et al. 1980: 45; Froeschner 1985: 49.
> *H. serratus* Heidemann 1901: 365 (not Fabricius).

Distribution. Endemic; Baltra, Isabela, Rábida, San Cristóbal, Santa Cruz, Santiago.
Bionomics. Phytophagous; arid to pampa and inversion zones; in moss and lichens on *Psidium guayaba*; February–June, December.

Tribe Niesthreini
Genus *Arhyssus* Stål 1870

A. species. New archipelago record.
Distribution. Indigenous?; Baltra, Floreana, Isabela, Marchena, Pinzón, Santa Cruz.
Bionomics. Phytophagous; arid to pampa zones; uv light; March–August.

Tribe Rhopalini
Genus *Liorhyssus* Stål 1870

L. hyalinus Fabricius 1794: 168 (*Lygaeus*); Barber 1925: 245; Schaefer et al. 1980: 45; Froeschner 1985: 50.
Corizus lugens Signoret 1859: 92 ; Barber 1934: 284.
Distribution. Indigenous?; cosmopolitan; Bartolomé, Daphne Major, Española, Floreana, Gardner at Floreana, Genovesa, Isabela, Marchena, Mosquera, Pinta, Pinzón, Rábida, Santa Cruz, Sombrero Chino, Wolf.
Bionomics. Phytophagous; arid to pampa zones; March–June.

Subfamily Serinethinae
Genus *Jadera* Stål 1860

J. silbergliedi Froescher 1985: 50
J. sanguinolenta Heidemann 1901: 365; (not Fabricius); Barber 1934: 284; Schaefer et al. 1980: 45.
Distribution. Endemic; San Cristóbal, Santa Cruz, Wolf.
Bionomics. Phytophagous; arid zone; host plant? *Caridiospermum* sp; at lights; January, March, June, December.

Family Stenocephalidae
Genus *Dicranocephalus* Hahn 1826

D. insularis (Dallas) 1852: 482; Heidemann 1901: 365; Barber 1925: 241, 1934: 282 (all *Stenocephalus*); Scudder 1957: 156; Lansbury 1965: 71; Froeschner 1985: 54; Moulet 1994: 362 (as *D. bianchii* (Jakovlev 1902)).
Distribution. Endemic; Genovesa, Isabela, Pinta, Santa Cruz, Santiago.
Bionomics. Phytophagous (on Euphorbiaceae?); littoral and arid zones; March–May.

Notes. The genus is otherwise Indo-African in distribution. Moulet (1994) synonomizes the species under *D. bianchii*, a widespread species of Africa and the Middle East. Schuh and Slater (1995) suggest that the species is conspecific with an (unnamed) African species, and that it was carried to the Galápagos by sailing ships. The problem with this idea of introduction is the infrequency of Galápagos visits by sailing ships from Old World ports, versus New World ports. Its early discovery (by Darwin in 1835) and presence on four islands, including Genovesa, which was off the routes of almost all sailing ships, also argues against a human mediated introduction. think the weight of the evidence suggests a natural colonization from the west, across the width of the Pacific, as improbable as this seems. The species remains scarce, but was present as recently as 1996.

Acknowledgements

G.G.E. Scudder (Univeristy of British Columbia, Vancouver), M.D. Schwartz (ECORC, Agriculture and Agri-Food Canada, Ottawa), and R.C. Froeschner (Smithsonian Institution, Washington, DC) were very gracious and helpful in identifying some of the new or problem species.

References

Anderson, N.M., and Polhemus, J.T. 1980. Four new genera of Mesoveliidae (Hemiptera, Gerromorpha) and the phylogeny and classification of the family. Entomol. Scand. **11**: 369–392.

Anderson, N.M., and Polhemus, J.T. 1972. The Lygaeidae of the Galápagos Islands (Hemiptera: Heteroptera). Proc. Calif. Acad. Sci., ser. 4, **39**: 87–103.

Ashlock, P.D. 1972. The Lygaeidae of the Galápagos Islands (Hemiptera: Heteroptera) Proceedings of the California Academy of Sciences Fourth Series. Vol. XXXIX, No. 88, pp. 87–103.

Ashlock, P.D., and Lattin, J.D. 1963. Stridulatory mechanisms in the Lygaeidae, with a new American genus of Orsillinae (Hemiptera: Heteroptera). Ann. Entomol. Soc. Am. **56**: 693–703.

Asquith, A. 1994. An unparsimonious origin for the Hawaiian Metrargini (Heteroptera: Lygaeidae). Ann. Entomol. Soc. Am. **87**: 207–213.

Bachmann, A.O. 1978. Heteroptera acuaticos de las Galápagos. Rec. Soc. Entomol. Argentina **37**: 131–135.

Baranowski, R.M., and Slater, J.A. 1997. A New World species of *Cymophyes* and a new species of *Xyonysius* from the Turks and Caicos Islands (Hemiptera: Lygaeidae). Fla. Entomol. **80**: 62–71.

Barber, H.G. 1925. Hemiptera-Heteroptera from the Williams Galápagos Expedition. Zoologica **5**: 241–254.

Barber, H.G. 1934. The Norwegian Zoological Expedition to the Galápagos Islands 1925, conducted by Alf Wollebaek. XI. Hemiptera-Heteroptera. Nyt Magazin for

Naturvidenskaberne **74**: 281–289. [Reprinted as Meddelesler fra det Zoologiske Museum, Oslo, nr. **42**, pp. 281–289 (1934).]

Barber, H.G. 1943. The *Halobates*. *Carnegie Institution of Washington, Publication*, **555**: 77–84.

Becker, P. 1992. Colonization of islands by carnivorous and herbivorous Heteroptera and Coleoptera: effects of island area, plant species richness, and 'extinction' rates. J. Biogeogr. **19**: 163–171.

Beebe, W.M. 1924. Galápagos Worlds End. C.P. Putman's Sons New York and London, 433 pp., illus.

Birch, M.C., Cheng, L., and Treherne, J.E. 1979. Distribution and environmental synchronization of the marine insect *Halobates robustus* in the Galápagos Islands. Proc. R. Soc. London (B) **206**: 33–52.

Brailovsky, A.H. 1985. Revision del genero *Anasa* Amyot-Serville (Hemiptera-Heteroptera-Coreidae-Coreinae-Coreini). Mono. Inst. Biol. Univ. Nat. Auton. Mexico **2**: 1–266.

Butler, A.G. 1877. Account of the Zoological Collection made during the visit of H.M.S. "Peterel" to the Galápagos Islands. X, Lepidoptera, Orthoptera, Hemiptera. Proc. Zool. Soc. London **1877**: 86–91.

Carvalho, J.C.M. 1976. Mirideos neoptropicais, CC: Revisao do genero *Horcias* Distant, com descricoes de especies novas (Hemiptera). Rev. Brasileira Biol. **36**(2): 429–472.

Carvalho, J.C.M., and Fontes, A.V. 1983. Mirideos Neotropicais, CCXXXIII: Genero *Dagbertus* Distant – Descricoes de especies e revisao das que ocorremna regiao (Hemiptera). Rev. Brasil. Biol. **43**: 157–176.

Carvalho, J.C.M., and Gagné, W.C. 1968. Miridae of the Galápagos Islands (Heteroptera). Proc. Calif. Acad. Sci., ser. 4, **36**: 147–219.

Carvalho, J.C.M., and Wagner, E. 1957. A world revision of the genus *Trigonotylus* Fieber. Arquivos do Museu Nacional, Brazil **43**: 121–155.

Champion, G.C. 1924. The Insects of the Galápagos Islands. Entomol. Mon. Mag. **60**: 259–260.

Cheng, L. 1997. Disjunct distributions of *Halobates* (Hemiptera: Gerridae) in the Pacific Ocean. Pac. Sci. **51**: 134–142.

Cheng, L., and Maxfield, L. 1980. Nymphs of two sea-skaters, *Halobates robustus* and *H. micans* (Heteroptera: Gerridae). Syst. Entomol. **5**: 43–47.

Cobben, R.H. 1965. A new shore-bug from Death Valley, California (Heteroptera: Saldidae). Pan-Pac. Entomol. **41**: 180–185. [*Pentacora sphacelata* recorded from Isla Santa Cruz (p. 181, figs. 2a–f).]

Dallas, W.S. 1851–1852. List of the specimens of Hemipterous Insects in the collection of the British Museum, London, Part I, pp. 1–368, 11 pls; Part II, pp. 369–592, 4 pls.

Doesburg, P.H., Van, Jr. 1968. A revision of the New World species of *Dysdercus* Guerin Meneville (Hemeroptera, Pyrrhocoridae) Zool. Verh. **97**: 1–215.

Dohrn, A. 1859. *Catalogus Hemipterorum*. iv–112 pp. Stettin: Herrcke and Lebeling.

Drake, C.J. 1954. Synonymic data and description of a new saldid (Hemiptera). Occ. Pap., Mus. Zool., Univ. Michigan, no. **553**, 5 pp.

Drake, C.J., and Chapman, H.C. 1953. Preliminary report on the Pleidae (Hemiptera) of the Americas. Proc. Biol. Soc. Wash. **66**: 53–60.

Drake, C.J., and Froeschner, R.C. 1967. Lace bugs of the Galápagos Archipelago (Hemiptera: Tingidae). Proc. Entomol. Soc. Wash. **69**: 82–91.

Drake, C.J. and Harris, H.M. 1946. A new mesoveliid from Ecuador (Hemiptera, Mesoveliidae). Bull. Brooklyn Entomol. Soc. **41**: 8–9.

Eger, J.E., II. 1978. Revision of the genus *Loxa* (Hemiptera: Pentatomidae). New York Entomol. Soc. **86**: 224–259.

Foster, W.A., and Treherne, J.E. 1980. Feeding, predation and aggregation behaviour in a marine insect (*Halobates robustus*: Hemiptera, Gerridae) in Galápagos Islands. Proc. R. Soc. London, B, **209**: 529–554.

Foster, W.A., and Treherne, J.E. 1982. Reproductive behavior of the ocean skater *Halobates robustus* (Hemiptera: Gerridae) in the Galápagos Islands. Oecologia, **55**: 202–207.

Froeschner, R.C. 1960. Cydnidae of the Western Hemisphere. Proc. U.S. Nat. Mus. **111**: 337–680.

Froeschner, R.C. 1968. Burrower bugs from the Galápagos Islands collected by the 1964 Expedition of the Galápagos Scientific Project (Hemiptera: Cydnidae). Proc. Entomol. Soc. Wash. **70**: 192.

Froeschner, R.C. 1976. Galápagos lace bugs: Zoogeographic notes and a new species of *Phatnoma* (Hemiptera: Tingidae). Proc. Entomol. Soc. Wash. **78**: 181–184.

Froeschner, R.C. 1980. Notes on a collection of Cydnidae (Hemiptera) from Ecuador and the Galápagos Islands made in 1964–1965 by the expedition of N. et J. Leleup. Bull. Inst. R. Sci. Nat. Belgique, Entomol. **52**: 1–4.

Froeschner, R.C. 1981. Heteroptera or true bugs of Ecuador; a partial catalogue. Smith. Contr. Zool. **322**: 1–147.

Froeschner, R.C. 1985. Synopsis of the Heteroptera or true bugs of the Galápagos Islands. Smith. Cont. Zool. **407**: 1–84.

Gerecke, R., Peck, S.B, and Pehofer, E. 1995. The invertebrate fauna of the inland waters of the Galápagos Archipelago, Ecuador; a limnological and zoogeographical survey. Arch. Hydrobiol./Supp. **107**: 113–147.

Göllner-Scheiding, V. 1978. Revision der Gattung *Harmostes* Burm. 1835 (Heteroptera, Rhopalidae) und einige bemerkungen zu den Rhopalinae. Mitteilungen aus dem Zool. Mus. Berlin **54**: 257–311.

Hamilton, S.W. 1983. *Neortholomus*, a new genus of Orsillini (Hemiptera-Heteroptera: Lygaeidae: Orsillinae). Univ. Kansas Sci-Bull. **52**: 197–234.

Heidemann, O. 1901. Papers from the Hopkins Stanford Galápagos Expedition, 1898–1899. I. Entomological Results (1): Hemiptera. Proc. Wash. Acad. Sci. **3**: 364–370.

Henry, T.J. 1997*a*. Monograph of the stilt bugs or Berytidae (Heteroptera), of the Western Hemisphere. Mem. Entomol. Soc. Wash. **19**: 1–148.

Henry, T.J. 1997*b*. Phylogenetic analysis of family groups within the infraorder Pentatomorpha (Hemiptera: Heteroptera), with emphasis on the Lygaeoidea. Ann. Entomol. Soc. Am. **90**: 275–301.

Henry, T.J., and Froeschner, R.C. (*Editors*). 1988. Catalog of the Heteroptera, or true bugs, of Canada and the continental United States. E.J. Brill, New York. 958 pp.

Herring, J.L. 1961. The genus *Halobates* (Hemiptera: Gerridae). Pac. Insects, **3**: 223–305.

Herring, J.L. 1966. The Anthocoridae of the Galápagos and Cocos islands (Hemiptera). Proc. Entomol. Soc. Wash. **68**: 127–130.

Hovarth, G. 1925. De Pentatomidarum genera *Loxa* Am. Serve. et de nova genere et affini. *Annales Musei Nationalis Hungarici*, **22**: 307–328, plates 4, 5.

Hurd, M.P. 1945. A monograph of the genus *Corythaica* Stål (Hemiptera: Tinigidae). Iowa State Coll. J. Sci. **20**: 79–99.

Kelton, L.A. 1955. Genera and subgenera of the *Lygus* complex (Hemiptera, Miridae). Can. Entomol. **87**: 277–301.

Kerzhner, I.M. 1968. Insects of the Galápagos Islands (Heteroptera, Nabidae). Proc. Calif. Acad. Sci., ser. 4, **36**: 85–91.

Lansbury, I. 1965. A review of the Stenocephalidae Dallas 1852 (Hemiptera-Heteroptera). Entomol. Mon. Mag. **101**: 52–92, 145–160.

Maldonado Capriles, J. 1990. Systematic catalogue of the Reduviidae of the world. Caribbean Journal of Science special edition, Univ. Puerto Rico, Mayaguez, Puerto Rico.

McAtee, W.L., and Malloch, J.R. 1925. Revision of the American bugs of the reduviid subfamily Ploiariinae. Proc. U.S. Natl. Mus. **67**: 1–153.

Moulet, P. 1994. Synonymies nouvelles dan la famille des Stenocephalidae Latreille, 1825 (Heteroptera, Stenocephalidae). Nouv. Rev. Entomol. (n.s.) **11**: 353–364.

Neiser, N. 1969. The Heteroptera of the Netherlands Antilles, VII. *Studies of the Fauna of Curaçoa and Other Caribean Islands*, **28**: 135–164.

Neiser, N. 1970. Corixidae of Suriname and the Amazon, with records of other Neotropical Species. *Studies on the Fauna of Suriname and Other Guyanas*, **12(45)**: 43–70.

Neiser, N. 1975. The Water Bugs (Heteroptera: Nepomorpha) of the Guyana Region. *Studies on the Fauna of Suriname and Other Guyanas*, **59**: 1-310, plates I-XXIV.

Parkin, P., Parkin, D.T., Ewing, A.W., and Ford, H.A. 1972. A Report on the Arthropods Collected by the Edinburgh University Galápagos Islands Expedition. 1968. Pan-Pac. Entomol. **48**(1): 100–107.

Polhemus, J.T. 1968a. A report on the Saldidae collected by the Galápagos International Scientific Project (Hemiptera). Proc. Entomol. Soc. Wash. **70**: 21–24.

Polhemus, J.T. 1968b. A new *Microvelia* from the Galápagos (Hemiptera: Veliidae). Proc. Entomol. Soc. Wash. **70**: 129–132.

Polhemus, J.T., and Chapman, H.C. 1979. Family Salididae; Shore Bugs *In* Menke, A., the semiaquatic and aquatic Hemiptera of California (Heteroptera: Hemiptera). Bulletin of the California Insect Survey, **21**: 16–33.

Polhemus, J.T., and Monzano, M. del Rosario. 1992. Marine heteroptera of the eastern tropical Pacific (Gelastocoridae, Gerridae, Mesoveliidae, Saldidae, Veliidae). *In* Insects of Panama and Mesoamerica. *Edited by* D. Quintero and A. Aiello. Oxford Univ. Press, New York. pp. 302–320.

Remane, R. 1964. Weiter Beitrage zur Kenntnis der Gattung *Nabis* Latreille (Hemiptera-Heteroptera, Nabidae). Zool. Beit., (n.f.), **10**: 253–314.

Rolston, L.H. 1983. A revision of the genus *Acrosternum* Fieber, subgenus *Chinaria* Orian, in the Western Hemisphere (Hemiptera: Pentatomidae). J. New York Entomol. Soc. **91**: 97–176.

Ruckes, H. 1957. Three new species of *Thyanta* Stål (Heteroptera: (Pentatomidae). Pan-Pac. Entomol. **33**: 175–180.

Sailer, R.L. 1948. The genus *Tricocorixa* (Corixidae, Hemiptera). Univ. Kansas Sci. Bull. **32**: 289–372.

Schaefer, C.W., Vagvolgi, J., and Ashlock, P.D. 1980. On a collection of Heteroptera (Hemiptera) from the Galápagos Islands. Pan-Pac. Entomol. **56**: 43–50.

Schuh, R.T. 1995. Plant bugs of the world (Insecta: Heteroptera: Miridae) systematic catalog, distributions, host list and bibliography. New York Soc., 1329 pp.

Schuh, R.T, and Schwartz, M.D. 1985. Revision of the plant bug genus *Rhinacloa* Reuter with a phylogenetic analysis (Hemiptera: Miridae) Bull. Am. Mus. Nat. Hist. **179**: 379–470.

Schuh, R.T., and Slater, J.A. 1995. True bugs of the world (Hemiptera: Heteroptera) classification and natural history. Comstock Publ. Assoc. Cornell Univ. Press, Ithaca, N.Y.

Scudder, G.G.E. 1957. The systematic position of *Dicranocephalus* Hahn, 1826, and its allies. Proc. R. Entomol. Soc. London, ser. A, **32**: 147–158.

Signoret, V. 1859. Monographie du genre *Corizus*. Ann. Soc. Entomol. Fr., ser. 3, **28**: 75–105.

Slater, J.A., and O'Donnell, J.E. 1995. A catalogue of the Lygaeidae of the world (1960–1994). New York Entomological Society, American Museum of Natural History, NY.

Stål, C. 1859. Hemiptera. Species novas descripsit. Kongliga Sevenska Fregatten Eugenies Resa Omkring Jorden, under befal af C.A. Vergin, aren 1851–3. Zoologi 1, Insecta, pp. 219–298.

Treherne, J.E., and Foster, W.A. 1980. The effects of group size on predator avoidance in a marine insect. Anim. Behav. **28**: 1119–1122.

Uhler, P.R. 1889. Hemiptera. *In* Scientific Results of Explorations by the U.S. Fish Commission Steamer Albatross. no. V.-Annotated Catalogue of the Insects collected in 1877–'88. *Edited by* L.O. Howard. Proc. U.S. Natl. Mus. **12**: 194.

Usinger, R.L. 1938. Biological Notes on the Pelagic Water Striders (*Halobates*) of the Hawaiian Islands, with Description of a New Species from Waikki (Gerridae, Hemiptera). Proceedings of the Hawaiian Entomol. Soc. **10(1)**: 77–84.

Usinger, R.L. 1941. The Present Status and Synonymy of Some Orsilline Species (Hemiptera, Lygaeidae). Bulletin of the Brooklyn Entomol. Soc. **36(3)**: 129–132.

Usinger, R.L., and Ashlock, P.D. 1966. Evolution of Orsilline Insect Faunas on Oceanic Islands (Hemiptera, Lygaeidae). *In* The Galápagos: Proc. Galápagos Inter. Sci. Proj. *Edited by* R.I. Bowman. pp. 233–235.

Van Duzee, E.P. 1933. The Templeton Crocker Expedition of the California Academy of Sciences, 1932. no. 4. Characters of twenty-four new species of Hemiptera from the Galápagos Islands and the coast and islands of Central America and Mexico. Proc. Calif. Acad. Sci. ser. 4, **21**: 25–40.

Van Duzee, E.P. 1937. No. 33. Hemiptera of the Templeton Crocker Expedition to Polynesia in 1934–1935. Proc. Calif. Acad. Sci. ser. 4, **22**: 111–126.

Villiers, A. 1970. Les Emesines des Iles Galápagos. Mission zoologique belge aux iles Galápagos at en Ecuador (N. et J. Leleup, 1964–1965), Resultats scientifique **2**: 227–237.

Villiers, A. 1978. Deux nouveaux Emesines des iles Galápagos. Bull. Soc. Entomol. Fr. **83**: 46–50.

Witlaczil, E. 1896. Die Ausbeute des "Pisani" an *Halobates* wahrend der Erdumseglung 1882–1885. Wien Entomol. Zeit. **5**: 177–182, 231–234.

Wygodzinsky, P.W. 1966. A monograph of the Emesinae (Reduviidae, Hemiptera). Bull. Am. Mus. Nat. Hist. **133**: 1–614.

Zheng, Le-Yi, and Slater, J.A. 1984. A revision of the genus *Pseudopachybrachius* (Lygaeidae: Hemiptera). Syst. Entomol. **9**: 95–115.

Chapter 21
Order Neuroptera

The Antlions and Lacewings

3 families, 5 genera, 8 species (5 endemic, 3 indigenous)
Minimally 8 natural colonizations

These insects are best characterized by their flattened and transparent wings which are filled with a net-like arrangement of veins and cross veins. The larvae of these are predators on other soil or plant dwelling insects. Sand-trap pits of antlion larvae are frequently seen in sheltered areas in fine, dry, loose soils of the arid zone of the islands. Larvae of green and brown lacewings feed on aphids or other small insects on vegetation. Adults may be predators, or may feed on pollen and nectar. These come to lights at night, sometimes in large numbers. The species composition and distribution of these insects in the Galápagos are now rather well known through the work of Klimaszewski et al. (1987) and Baert et al. (1992). Most of my specimens of this order have been placed in the collections of the California Academy of Science, San Francisco. Others are in the USNM. O. Flint identified the record of *Chrysoperla externa* (Hagen).

Key to Neuroptera of the Galápagos Archipelago

1a Antenna knobbed and short, no longer than length of head and thorax combined; large specimens (body length 20–25 mm, wingspan 45–55 mm) with long and slender body and elongate wings; Family Myrmeleontidae 7
1b Antenna evenly narrowly elongate, distinctly longer than head and thorax combined; smaller specimens (body length 4.0–7.0 mm, wingspan 7.0–24.0 mm) with body shorter and wings either approximately oval or slightly angular in shape 2
2a Small or medium size, dark brown or brownish insects (body length 4.0–6.0 mm, wingspan 7.0–13.0 mm); with setose and broad or slender oval wings; radius of forewing with two or more branches; Family Hemerobiidae 6
2b Medium size, whitish, yellowish, or greenish insects (body length 5.0–7.0 mm, wingspan 20–24 mm); with less setose and slightly angular wings; radius of forewing with one branch, radial sector with zigzag appearance; Family Chrysopidae 3

3a Head with a black spot on each cheek under eye; palpi mostly black; venation pale, marked with black with some cross veins either wholly of partly dark 5

3b Head uniformly yellowish, without black spots; palpi uniformly yellowish; venation uniformly yellowish, sometimes some cross veins appearing darker 4

4a Dorsal surface of first antennal segment and lateral margin of pronotum uniformly as pale in colour as adjacent areas. . . . *Ceraeochrysa cincta*

4b Dorsal surface of first antennal segment and lateral margin of pronotum with reddish brown stripe . . . . *Chrysoperla externa*

5a Body greenish, much marked with black; forewing with stigma bearing at least 3 dark spots, gradate veins divergent, with about 4 veins in inner and 7 in outer sector and with inner gradate veins closer to the radial sector than to outer gradates . . . . *Crysopodes nigripilosa*

5b Body yellowish; forewing with stigma uniformly greenish, gradate veins in parallel series, with 6 or 7 veins in each sector, and with inner gradates almost as close to outer gradates or to the radial sector
. . . . *Chrysoperla galapagoensis*

6a Larger, wing length about 6 mm; head and thorax dorsally uniformly dark; forewing broadly oval, widest near the base, with costal area abruptly widened basally; radial sector with at least 3 branches, all being connected with R_1 + Rs combined; media anterior with first fork in vicinity of radial sector; veins with dark and paler sectors but not dotted.. . . . . *Megalomus darwini*

6b Smaller, wing length about 4 mm; head and thorax dorsally pale with distinct dark dots; forewing slender oval, widest in apical third, with costal area moderately narrow and slightly broadened basally, radial sector with two branches, basal one being connected with media anterior; veins darkly dotted . . . . *Sympherobius barberi*

7a Wings with membrane transparent and clear, almost evenly, narrowly elongate and only insignificantly broadened in apical fifth, female tergum 10 with stout spines clustered at the bottom *Myrmeleon perpilosus*

7b Wings with membrane maculated, strongly broadening apically and widest in apical fifth; female tergum 10 with spines or setae more evenly 8

8a Forewing with membrane suffused near vein junctions and with rather small dark brown spots at the origin Rs + MA and its forks, cubital fork, end of CuP + 1A and before stigma; hind wing with veins mostly dark brown and little membrane suffusion (for illustrations and details see Stange (1969). . . . . *Galapagoleon darwini*

8b Forewing with membrane bearing pale and dark large spots often forming distinct bands, and with one black apical spot preceded by two pale ones; hind wing with several large spots connected through the middle (this species is erroneously recorded from Galápagos)

. . . . *Millerleon subdolus* (Walker) (= *Dimares formosus* Banks)

Superfamily Myrmeleontoidea
Family Myrmeleontidae
The Antlions

Millerleon subdolus (Walker), under the name *Dimares formosus* Banks (1908: 31) has been erroneously reported from the Galápagos (Esben-Peterson 1920: 191; Navás 1912: 229; Stange 1969: 189). It is not listed here (see Klimaszewski et al. 1987; Stange 1989: 460).

Genus *Galapagoleon* Stange 1994 (endemic genus)

G. darwini (Stange) 1969: 190 *(Brachynemurus)*; Parkin et al. 1972: 103; Baert et al. 1992: 143; Stange 1994: 97.

Distribution. Endemic; Baltra, Española, Fernandina, Floreana, Genovesa, Isabela, Marchena, Pinta, Pinzón, Santiago, Santa Cruz, Santa Fé, Seymour, Wolf.

Bionomics. Predator; arid zone; January, March–May. Maximum flight activity seems to be about 1900 h (Baert et al. 1992).

Genus *Myrmeleon* Linnaeus 1767

M. perpilosus Banks 1924: 177, Stange 1969: 196; Parkin et al. 1972: 102; Klimaszewski et al. 1987: 3039; Baert et al. 1992: 144.

Distribution. Indigenous; Mexico to Peru; Baltra, Campeón, Española, Fernandina, Floreana, Gardner at Floreana, Genovesa, Isabela (Sierra Negra, Wolf, Alcedo, Cerro Azul), Marchena, Pinta, Pinzón, Rábida, San Cristóbal, Santa Cruz (type locality Conway Bay), Santa Fé, Santiago, Wenman.

Bionomics. Predator; arid zone mostly, to pampa zone; January–June; at lights, malaise traps.

Superfamily Hemerobioidea
Family Hemerobiidae
The Brown Lacewings
Genus *Megalomus* Rambur 1842

M. darwini Banks 1924: 179; Parkin et al. 1972: 102; Klimaszewski et al. 1987: 3035; Baert et al. 1993.
  Distribution. Endemic; Fernandina, Floreana, Isabela (Alcedo, Cerro Azul, Sierra Negra), Rábida, Santiago, San Cristóbal, Santa Cruz, Seymour.
  Bionomics. Predator; arid zone (but higher on occasion); January–June; at lights, malaise traps.

Genus *Sympherobius* Banks 1905

S. barberi (Banks) 1903: 241 (*Hemerobius*); Klimaszewski et al. 1987: 3035; Oswald 1988: 427; Baert et al. 1992: 146.
  Distribution. Indigenous; most of USA to Peru, Hawaii (adventive) and Revillagigedo Islands; Fernandina, Isabela (Cerro Azul, Alcedo, Sierra Negra), Rábida, Santa Cruz, Santiago, Seymour.
  Bionomics. Predator; arid and culture zones, *Miconia* zone; January–June; at lights, malaise traps.

Family Chrysopidae
The Common or Green Lacewings
Genus *Chrysoperla* Steinmann 1964

C. galapagoensis (Banks) 1924: 179 (*Chrysopa*); Parkin et al. 1972: 102 (as near *C. galapagoensis*); Brooks and Barnard 1990: 271; Baert et al. 1992: 144.
  Distribution. Endemic; Baltra, Española, Fernandina, Floreana, Genovesa, Isabela (Cerro Azul, Alcedo), Pinzón, Rábida, Santiago, Santa Cruz, Wolf.
  Bionomics. Predator; arid to humid forest zones, higher on occasion; March–June, December; malaise traps and at lights.

C. externa (Hagen): Baert et al. 1992: 145. (as "new species")
  Distribution. Indigenous? Floreana, San Cristóbal.
  Bionomics. Humid forest and pampa zones; M.V. lights, sweeping, FIT; January–March.

Genus *Ceraeochrysa* Adams 1982

C. cincta (Schneider) 1851(*Chrysopa*); Brooks and Barnard 1990: 268; Baert et al. 1992: 145.
  C. wollebaeki Esben-Peterson 1934: 291; Klimaszewski et al. 1987: 3034 (*Chrysopa*)

Distribution. Endemic; Fernandina, Floreana, Isabela (Alcedo, Cerro Azul, Sierra Negra), San Cristóbal, Santa Cruz.

Bionomics. Predator; arid to humid forest (or agricultural) zones; January–April; at lights, malaise traps.

Genus *Chrysopodes* Navás 1913

C. nigripilosa (Banks) 1924: 177(*Chrysopa*); Parkin et al. 1972: 102; Brooks and Barnard 1990: 272; Baert et al. 1992: 145 (*Neosuarius*).

Distribution. Endemic; Baltra, Fernandina, Floreana, Isabela (Darwin, Alcedo, Sierra Negra, Wolf), Rábida, Santa Fé, Santiago, Santa Cruz, Seymour.

Bionomics. Predator; arid zone, higher on occasion; January–May; at lights.

References

Baert, L., Desender, K., and Peck, S.B. 1992. New data on the Neuroptera of the Galápagos Islands, Ecuador. Bull. Inst. Rot. Sci. Nat. Belgique, Entomol. **62**: 143–147.

Banks, N. 1924. Neuroptera from the Williams Galápagos Expedition. Zoologica, **5**: 117–180.

Brooks, S.J., and Barnard, P.C. 1990. The green lacewings of the world: a generic review (Neuroptera: Chrysopidae). Bull. Br. Mus. (Nat. Hist.) Entomol. **59**: 117–286.

Esben-Petersen, P. 1920. Revision of some of the type specimens of Myrmeleonidae, described by Navas and placed in the Vienna Museum. Ann. Soc. Entomol. Belgique, **60**: 190–196.

Esben-Petersen, P. 1934. The Norwegian Zoological Expedition to the Galápagos Islands 1925, conducted by Alf Wolleback. XII. Neuropterous insects from Galápagos Islands. Nyt Magazin Naturvidenskaberne, vol. 74, pp. 291–294. [Reprinted as Meddelelser fra det Zoologiske Museum, Oslo, nr. 43, pp. 291–294 (1934).]

Klimaszewski, J., Kevan, D.K. McE., and Peck, S.B. 1987. A review of the Neuroptera of the Galápagos Islands, with a new record for *Sympherobius barberi* (Banks) (Hemerobiidae). Can. J. Zool. **65**: 3032–3040.

McLachlan, R. 1877. Account of the zoological collection made during the visit of H.M.S. *Peterel* to the Galápagos Islands. Neuroptera. Proc. Zool. Soc. London, **1877**: 84–86.

Navas, R.P. 1912. Myrmeleonides (Ins. Neur.) noveaux ou peu connus. Ann. Soc. Sci. Bruxelles, **36**: 203–248.

Oswald, J.D. 1988. A revision of the genus *Sympherobius* Banks (Neuroptera: Hemerobiidae) of America north of Mexico with a synonymical list of the world species. J. N.Y. Entomol. Soc. **96**: 390–451.

Parkin, P., Parkin, D.T., Ewing, A.W., and Ford, H.A. 1972. A report on the arthropods collected by the Edinburgh University Galápagos Expedition, 1968. Pan-Pac. Entomol. **48**: 100–107.

Stange, L.A. 1969. Myrmeleontidae of the Galápagos Islands (Insecta: Neuroptera). Acta Zool. Lilloana **25**: 187–198, Figs. 1–15.

Stange, L. 1989. Review of the New World Dimarini with the description of a new genus from Peru (Neuroptera: Myrmeleontidae). Fla. Entomol. **72**: 450–461.

Stange, L. 1994. Reclassification of the New World antlion genera formerly included in the tribe Brachynemurini (Myrmelentidae). Insecta Mundi **8**: 67–120.

Chapter 22
Order Strepsiptera

The Twisted-wing Insects
1 family, 1 genus, 1 species (1 indigenous)
Minimally 1 natural colonization

Strepsiptera are enigmatic insects whose larvae and adult females are endoparasitic in a variety of other insect hosts and whose adult males are free flying. The mobile larvae leave the body of the larvaform adult female strepsipteran and then the body of the host insect when the host is in a suitable place. Then the larvae find and board a new host. This may be on a flower for the species that infect bees or wasps. In the Galápagos it is probably on vegetation inhabited by the host plant feeding leafhoppers or bugs. This order is probably an ancient sister lineage to the beetles. A review of Neotropical Strepsiptera is that of Kathirithamby (1992).

Family Elenchidae
Genus *Elenchus* Curtis 1831

E. koebeli (Pierce) 1908: 81; Peck and Peck 1989: 203.
    Distribution. Indigenous; widespread in North, Central and South America; Isabela, San Cristóbal, Santa Cruz
    Bionomics. Larvae are parasites, probably in planthoppers (Delphacidae, Homoptera) and adult females remain in the host. They have been reared in the Galápagos on the delphacid bugs *Nesosydne olipor* and N. *alcmaeon* (Abedrabbo et al. 1990: 125); transition and humid forest zones. *Elenchus koebeli* probably dispersed to the Galápagos as larvae or an inseminated female in a host, because the adult females are flightless, and the males live only a short time, maybe for only 24 h. Collected in January to March at ultraviolet light traps or flight-intercept traps, or in infected host insects. Leafhoppers (Cicadellidae) found with strepsiptera in Galápagos are *Balclutha lucida* and *B. rosea*.

References

Abedrabbo, S., Kathirithamby, J., and Olmi, M. 1990. Contribution to the knowledge of the Elenchidae (Strepsiptera) and Dryinidae (Hymenoptera: Chrysidoidea) of the Galápagos Islands. Bull. Inst. Entomol. "Guido Grandi" Univ. Bologna **45**: 121–128.

Kathirithamby, J. 1992. Strepsiptera of Panama and Mesoamerica. *In* Insects of Panama and Mesoamerica. Selected Studies. *Edited by* D. Quintero and A. Aiello. Oxford University Press, Oxford, U.K. pp. 421–431.

Peck, S.B., and Peck, J. 1989. *Elenchus koebeli* (Pierce) (Elenchidae): First record of Strepsiptera from the Galápagos Islands, Ecuador. Coleopt. Bull. **43**: 203–204.

Chapter 23
Order Siphonaptera

The Fleas

3 families, 4 genera, 4 species (1 endemic, 3 adventive)
Minimally 1 natural colonization

Fleas are secondarily wingless insects that have evolved from a fly- or scorpionfly-like ancestor. The larvae are scavengers in the nests or burrows of their bird or mammal hosts, and adult fleas feed on the blood of their hosts. Sea mammals and other mammals that do not have dens or nests usually do not have fleas. Only one species seems to naturally occur in the Galápagos. More species may be found in the nests or burrows of various seabirds. No adult fleas were found in our search of the plumage of 590 individual land and sea birds. There may have been fleas on the now extinct genera and species of endemic rice rats. We may never know.

Family Rhopalopsyllidae
Genus *Parapsyllus* Enderlein 1903

P. cedei Smit 1970: 244; 1987: 145.
Distribution. Endemic; Genovesa, Plaza, Santa Cruz.
Bionomics. January; hosts are Audubon's shearwater (*Puffinus lherminieri*) and Madeiran and Galápagos storm petrels (*Oceanodroma castro* and *O. tethys*). This flea is found in the nests of these birds.

Family Tungidae
Genus *Tunga* Jarocki 1838

T. penetrans Linnaeus 1758 page; Conway and Conway 1947: 200; Treherne 1983: 54. The Chigoe flea.
Distribution. Adventive; circumtropical; Santiago, Floreana.
Bionomics. February–March; locally called "*niguas*," these fleas burrow deep into skin of humans (often under the toenails) or livestock like pigs and can cause considerable pain.

Family Pulicidae
Genus *Ctenocephalides* Stiles and Collins 1930

C. felis (Bouché) 1835: 505. New archipelago record.
Distribution. Adventive; cosmopolitan; Santa Cruz, Santiago.

Bionomics. The cat flea. More records should be expected on more islands from domestic and feral cats; the only known specimens to date were captured in pit trap in the Santa Cruz *Miconia* zone, March 1989; and on dogs on Santiago, and Santa Cruz; in February and March.

Genus *Pulex* Linnaeus 1758

P. species (prob. *irritans* or *simulans*); New archipelago record.
Distribution. Adventive; cosmopolitan; Isla Santiago, Aguacata Camp
Bionomics. The record was taken on a dog in March.

Acknowledgements

A.H. Benton (Fredonia, NY) provided the cat flea identification.

References

Conway, A., and Conway, F. 1947. The Enchanted Islands. G.P. Putnam's Sons, New York, 280 pp.

Smit, F.G.A.M. 1987. An illustrated catalogue of the Rothschild collection of fleas (Siphonaptera) in the British Museum. vol. 7, Malacopsylloidea. (Malacopsyllidae and Rhapalopsyllidae). Oxford Univ. Press.

Smit, F.G.A.M. 1970. A new species of flea from the Galápagos Islands. Entomol. Ber. **30**: 244–247, figs. 1–3.

Theherne, J. 1983. The Galápagos Affair. Triad Grafton, London. 269 pp. (paperback).